THE GRAMMAR OF TELEVISION PRODUCTION

THE GRAMMAR OF
Television Production

DESMOND DAVIS

Revised by
JOHN ELLIOT

Further revised by
MIKE WOOLLER

Drawings by
FRANK WHITE
MICHAEL KNIGHT

Additional drawings
by VIC SYMONDS

PUBLISHED UNDER
THE AUSPICES OF
THE SOCIETY OF
FILM AND TELEVISION
ARTS
BY
BARRIE AND JENKINS
LONDON

First published 1960 by
Barrie and Rockliff
(Barrie & Jenkins Ltd)
24 Highbury Crescent
London, N5 1RX
Second impression 1961
Second edition revised and re-set 1966
Fourth impression 1969
Third edition revised 1974
Printed and bound in Great Britain by
REDWOOD BURN LIMITED
Trowbridge & Esher

THE SOCIETY OF FILM AND TELEVISION ARTS

President: H.R.H. The Princess Anne
Vice President: Richard Attenborough C.B.E.

SFTA is a Society where people from film and television can rub shoulders socially and compare notes professionally. Until 1959, people who worked in films joined the British Film Academy, and those in television joined the Guild of Television Producers and Directors. But the film people and the television people realised that they had quite a lot in common. So they merged. And the Society of Film and Television Arts was born. Through this distinguished lineage the Society can look back on many years of involvement in and awareness of the development of films and television.

Its membership includes many who have made outstanding contributions to that development, and many others who have yet to make their mark.

SFTA was formed to cultivate among its members the highest possible standards in film and television production, both artistically and technically. Every year the Society celebrates the works that epitomise these ideals, and the annual Awards presentation at the Albert Hall is perhaps the best-known of its activities. The Awards have achieved an importance comparable with the Hollywood 'Oscar', and their presentation has become a major event shared by television viewers all over Britain.

During each year the Society arranges private viewings of the best new feature films and others not generally available. Members are admitted free, and can usually bring a guest. The Society also holds meetings at which important film and television matters are discussed.

In addition, the Society publishes a quarterly JOURNAL devoted to film and television subjects, and a Members'

Newsletter which gives news of SFTA activities and events within both industries.

At the Society's headquarters in Piccadilly there are club facilities, meeting rooms and two of the finest viewing theatres in London. There is also a library of books on film and television subjects, as well as a valuable collection of British film scripts. The Society hopes to widen these facilities in the near future, and plans for a new building to house them are currently being considered.

The Society is recognised by Government Departments as a source of information, and is increasingly being consulted to submit evidence to Governmental and other Committees of Inquiry.

SFTA is a registered charity. Its first President was HRH The Prince Philip, succeeded in 1966 by The Earl Mountbatten of Burma. He was succeeded in 1972 by H.R.H. The Princess Anne. Its Vice-President is Richard Attenborough, C.B.E.

The Society is governed by a 20-strong Council of Management, elected annually.

INTRODUCTION

"The Grammar of Television Production" is the work of a distinguished director, the late Desmond Davis, who died tragically at the time of its original publication. In bringing it up to date I have kept the greater part of his text, since after a long and critical look, I cannot see how most of it could be put better. Because television has changed rapidly during the past few years, I have felt bound to make some additions and to suggest a slight change of attitude; but although fashions have altered, the vocabulary has grown and techniques have become more sophisticated, the basic grammar of the language remains as Desmond Davis set it down.

He was at pains, in his first edition, to point out that it was indeed only grammar—a set of rules governing fundamental technique—and that sound grammar is only a foundation to imaginative creative work, not a substitute for it. At the same time, rules were laid down absolutely and arbitrarily as precepts which the good practitioner must accept totally in his mind however much, in practice, he departs from them.

Not everyone accepts this now, and I would ask the newcomer to take what follows not as holy writ but as an enormously useful guide to the established and tested conventions of the craft. It is full of common-sense and the harvest of many people's experience, and should not be ignored.

During the past decade, television has been busy writing its own shorthand. Originally it was very much a longhand script: one saw things as they happened, and at the speed at which they happened—the speed of life. The theatre has never been bound by this, nor the film, nor the novel; they all compress, take flight, exaggerate, hop skip and jump over time and space as their theme requires. Television has been learning to do this.

Perhaps the greatest change in technique is that camerawork has grown more flexible as television cameras have become increasingly mobile and sensitive. The accepted style of camera direction is quicker and lighter than it used to be, and dynamic following of the subject often more important than the conventional set-up.

Another major development has been an enormous advance in the possibilities of pre-recording of and filming for programmes. Although this should not alter the director's approach to his individual shots, it does give him whole new areas of control over his production, and a fresh attitude to its shape, tempo and punctuation.

Consequently, I have modified what the original edition called 'The Conventions', and added chapters on recording and modern studio equipment, and a few other notes which I hope will be useful. My thanks are due to all those colleagues who have given help and advice.

JOHN ELLIOT

Note to Third Edition

To take account of subsequent developments in this field, the present, 1974, printing of this book has been further revised by Mike Wooller.

A CODE FOR TELEVISION PRODUCTION

Every art and craft, over the years, gathers for itself a lore, a collection of rule of thumb 'do's' and 'don'ts' which are handed to the newcomer; 'do's' and 'don'ts' born of long experience and experimentation which form a short cut to his ultimate proficiency.

Much contained in the book is opinion and not incontrovertible fact, but it is informed opinion and that held by a large majority of the most experienced television producers and directors in this country, the country that has the longest history of television of any in the world. At least, the beginner who follows the precepts set out in this book will avoid the grosser errors of television production. Perfection in the technique of presentation can never take the place of intrinsic merit in the content, but content is conditioned by presentation and cannot exist independently.

RULES AND ART

It is reasonably true to say that the television director can never see his production unless it is recorded. In the control room he is too busy with the future to criticize the past. He never has the time to try something a different way and quietly assess which is best. It is therefore particularly difficult for him to build up hard and fast 'do's' and 'don'ts'.

But every art *has* got its rules. The masterpiece of literature would surely lose something if it were mis-spelt and wrongly punctuated. In point of fact, bad spelling and punctuation would do exactly what bad camera work would do; it would draw attention to itself and away from the subject matter.

However, this is a book not so much on the art of television, as on the technique; the second most important thing in television production and direction. The first things are taste, talent and imagination which cannot be governed by rules; whereas technique, which can never be a substitute for them, can.

It may well be conceded that in no art can there be rules so rigid and inflexible that they can never, in any circumstances, be broken, but before they are broken they should be known and they should only be broken after the most careful consideration and for very good and special reasons.

THE AIM OF TECHNIQUE

As this book is solely concerned with technique, it may give the impression that only technique is important. This, of course, is the reverse of the case.

The rules governing camera work and editing have one main object: to make the camera work and editing as unnoticeable as possible. While never sacrificing dramatic or pictorial effect, they work to eliminate everything in the technique that may be ugly, jarring, ridiculous, obtrusive or visually discordant. They aim to make the camera work and the editing invisible to the beholder so that nothing may come between him and his concentration on the subject matter.

Content, subject matter, the writing, the performance, the acting; these are the things that matter most in all media of communication or entertainment. The television producer must never allow his mind to become so filled with considerations of method and technique that those of content and interpretation are forgotten.

The purpose of this book is so to smooth out technique that it will never distract. The best technique highlights that which is important, eliminates that which is unimportant and obliterates itself by its own perfection.

THE CODE

The Code consists of three main ingredients:

1. **Conventions.** These should be accepted and adhered to by the producer and director for the good and convenience of himself, his fellows and the public.

2. **Rules.** These may be regarded as binding and should only be broken in exceptional circumstances, to achieve some special surprising, comic or bizarre effect.

3. **Hints and tips.** These are proved and useful rule-of-thumb guides; they may be regarded as less binding than the rules, but they should only be broken by the experienced producer or director and then only for some good and cogent reason.

The Rules and the Hints and Tips are gathered under seven headings: Editing, Lenses, Composition, Camera work, General, Equipment and Recording. The Rules are numbered and are printed in heavy type. The Hints and Tips are numbered and lettered and are printed in heavy type. Both the Rules and Hints and Tips are discussed and explained in normal type. At the end of the book there is a glossary of common technical terms.

Note. The roles of producer and director vary from time to time, and from studio to studio. Here the terms 'producer' and 'director' normally refer to the person actually controlling the cameras, who nowadays is more usually known as the 'director'.

CONVENTIONS

1. **THE CUT. This is the normal method of getting from one camera to another when the action is continuous in time. It may also be used as a means of moving from scene to scene.**

2. **THE DISSOLVE OR MIX. This is normally used when a lapse of time is to be indicated or in getting from one scene to another between which there could have been a lapse of time or in getting to and from captions. It is not, in normal circumstances, used when the action is continuous.**

3. **THE FADE IN AND OUT. The fade out to black and the fade in from black are used in the same way as the dissolve or mix, but are accepted to mean an even longer lapse of time and, as it were, to indicate the end of a chapter.**

Notes on 1, 2, *and* 3. These are time-honoured and internationally observed conventions, which have always been understood and accepted by the public. They constitute an invaluable instrument in our hands. If we abuse or misuse them we confuse the public, destroy the meaning of the conventions and do ourselves and our colleagues a disservice. They are the grammar and punctuation of television.

(The rule governing the use of the dissolve—or mix—may sometimes be broken successfully; e.g. in a music or dream sequence where the accepted canons of time and place do not apply; or in an Outside Broadcast of a public event when a director wishes to convey a change of mood in the course of continuous shooting. But these are special circumstances.)

4. **RIGHT AND LEFT. In the studio, the rehearsal room, and on scripts, right and left shall mean camera right and left—that is right and left looking from the camera towards the subject.**

This may well seem to many to be too obvious to be worth stating but the convention of the theatre has always been to refer to right and left as actor's right and left. A newcomer from the theatre might, therefore, be confused with the different system.

RULES—AND HINTS AND TIPS

EDITING

1. When fading up opening caption routines at the start of a programme, never allow the picture to precede the sound; bring them up both together if possible but, if not, contrive for the sound to precede the picture by the tiniest fraction.
It is difficult to give an exact reason for this rule but this does not make the rule less valid. A picture without sound is dead and unpleasing, sound without a picture is tolerable and often pleasant.

1a. Roll captions at a speed at which they can comfortably be read aloud.
Captions which move too slowly can be an annoyance. Captions that move too fast are infuriating; they cannot be read and they tend to cause strobing, which is ugly. The speed at which they can be read aloud is the correct speed.

1b. Never have a caption on the screen that says one thing while a voice is saying something else. Sound and picture should, at all times, be partners and never rivals.
The eye and the ear cannot take in two conflicting pieces of information at once.

For this reason, the greatest care must be taken in writing, editing and timing a commentary. The commentator should always be speaking of what is on the screen or of what is just about to come on to it. For instance, the commentator should not be saying: 'At this time, Mr. Baldwin, at ten Downing Street, was in the process of re-shuffling his cabinet' when, on the screen, we see Ramsay MacDonald addressing a meeting at Lossiemouth. Let the sound and the picture always work in harmony. Because words take time to be understood, it is often necessary for them slightly to precede the picture, which can be understood at once if we know what we are supposed to see. When the visual interest of one picture has been exhausted, the viewer can pay attention to words preparing the way to the next picture, but they must not be too specific. In the example given, Baldwin's name should immediately precede his picture.

1c. In superimposing a caption on a background picture, make sure that the lettering and the background are of contrasting tones. When working in colour it is also necessary to make sure that lettering and background will have contrast in monochrome reception.

This may seem too obvious to be worth stating but it is, nevertheless, frequently overlooked. Use light letters on a dark background and dark letters on a light background, otherwise they will disappear. Avoid backgrounds that are half their area light and half their area dark unless light letters are confined to the dark area or dark letters are confined to the light area. For roller captions these mixed backgrounds must always be wrong.

2. Cut, dissolve (mix) and fade, always to the rhythm of the music, never against it.

The reasons for this rule should be self-evident but it is one that is often broken. It is obviously jarring to have the music and the picture operating at conflicting tempi, the punctuation of one not coinciding with that of the other.

On rhythmic music the cut should invariably come at the end of a musical phrase. On any music it should coincide with the musical punctuation.

It is up to the director, so to govern the action that the correct place to cut musically is also the correct place to cut visually.

3. Fade out music only at the end of a musical cadence, never in the middle.

It is extremely irritating to the ear for music to stop before the cadence has been completed.

The two exceptions to this rule are:

(1) Where the music is faded out so gently and gradually under dialogue or some other sound that one is not aware of its going or its end.

(2) Where the music is immediately swamped by another more forceful and much louder sound.

4. Avoid an ugly and meaningless fast dissolve (mix). Accept two seconds as the minimum and three seconds as the standard.
Except, possibly, in getting from one caption to another, the fast dissolve or mix has little meaning. It does not indicate a lapse of time, which is the only reason for the existence of any sort of dissolve, and the fast dissolve merely looks like an untidy cut.

5. Never cut between moving cameras, particularly panning cameras, or between one camera that is moving and one that is stationary.
A cut between moving cameras has a most unpleasant effect on the eye. It is ugly and disturbing and makes the cut very noticeable.

A permissible exception to this rule is when both cameras are panning in the same direction at the same speed. It would be permissible, for instance, to cut from a panning shot of a car going along a road to a similar shot of another car travelling at the same speed in the same direction, or to cut from a long shot of a travelling car to a close shot of the same car. The point here is that in this instance the camera is virtually not panning. It is only panning in relation to the background, which is unimportant and hardly noticed, and is stationary in relation to the important thing, the car.

Another possible exception is in cutting *to* a panning camera on fast-moving action. This tends to accelerate the tempo and increase the excitement. However, it should only be done in these special circumstances. Never, in any circumstances, is it permissible to cut *away from* a panning camera. This is intolerable save in the exception set out in the paragraph above.

6. Do not dissolve (mix) between moving cameras, particularly panning cameras, or from one static camera to a moving one and vice versa.

To dissolve between moving cameras has a very ugly effect, it tends to make the viewer feel slightly sick. The faster the motion, the more sick it makes him feel.

The same legitimate exception exists for dissolving as the first one laid down for cutting (see rule 5 above) and for the same reason. It is permissible when both cameras are moving in the same direction at the same, very steady speed.

This is a rule that is often broken in feature film making. The larger screen seems to make it less unpleasant and the pan or other movement is always rather slow and very steady, a difficult thing to achieve in the hurly-burly of 'live' television. It is still felt by many to be, artistically, a doubtful manœuvre.

Though it is widely held that a dissolve should only take place between static cameras, a dissolve should preferably be animated, i.e. have movement in it. It is better, however, to do this by moving the subject than by moving the frame.

7. Always, when possible, cut on movement within the frame; cut when the subject is in the act of sitting, rising, turning, rather than when the subject is stationary. Even in close-up, prefer a moment for cutting when the head is in motion.

Movement of the subject makes the cut less noticeable, in fact virtually invisible. Cuts between static subjects always tend to be more obtrusive than those between ones in motion.

Timing is all-important. Frequently one sees cuts that are made either too early, just before the move has started, or too late, just after it has finished. Part of the move should be seen in both shots.

If, for instance, you have a long shot of a man walking to a chair and sitting, whereon you wish to cut to a close shot of him, you should see him start to sit in the long shot and see him complete the sitting movement in the close shot. Similarly, on rising, you should see him start to rise in the close shot and finish rising in the longer shot. You should *not* cut to the longer shot *before* he rises and both cameras should be *static* on the cut. They should not pan him into or out of, the chair. If they do this you will be cutting on a pan, which is ugly.

Another motivation for the cut which is also movement to a limited extent, is the 'LOOK'. A man is sitting at a desk. We hear a 'click', he looks up. We cut to what he sees—someone coming in at the door. Although in this case, the movement is very small, possibly, only the shift of the eye-line, the cut is so palpably motivated, it so exactly satisfies our instinct at that moment that it will be unnoticed.

Where, on dialogue sequences for instance, movement cannot be the governing factor or where there is virtually no movement, then cut purely for dramatic reasons, to favour the person who is speaking or for reaction or expected reaction. However, if you can cut on movement or a 'Look', do so, it will make the cut very much less noticeable.

7a. When cutting on dialogue do not be rigidly bound by ends of speeches.

When cutting on action, the rule, wherever possible, to cut on movement, should apply. When cutting on dialogue, as between a pair of matched and complementary close-ups or a pair of over-shoulders, there is little or no movement and another guide must be found. The guide is a simple one—favour the subject that is most important at any given moment. Generally, the person who is speaking is most important but this is not always the case. Sometimes the speaker is less important than the reaction to what is being said. In this case use the camera that is favouring the person who is listening.

There is a limit to the speed of cutting which the eye will tolerate. Do not, therefore, try and cut back and forth for every syllable that is uttered. A few odd words spoken off camera do not matter. Always favour the person who is imparting the most information whether it is by words or facial expression.

8. Never change the picture by cutting, dissolving, panning or tracking without adequate motivation from either the action or the accompanying sound.

Any change of picture tends to attract the attention of the audience away from the subject matter and towards the technique, therefore never change it unless the next picture says something different, that *has* to be said, something that emphasizes a point or adds to the comprehension of the audience. *Never change a picture just for the sake of change; it is a mere distraction.*

8a. Do not imagine that an illegitimate dissolve (mix) will conceal an illegitimate cut.

It is a fallacy that a bad and jarring cut will be concealed by mixing. This merely piles error upon error. If it is a bad cut, it will be an even worse mix. If the action is continuous in time then it *should* be a cut. If there is a change of scene in which there is a lapse of time to be indicated or in which there *could* have been a lapse of time, then it *could* be a mix. Otherwise, the rules are rigid. If the cut is bad, alter the position of the cameras or the timing of the cut. Do not imagine that it can be smudged over and will not be noticed.

Remember that a mix is much more noticeable than a cut. It is dead, blurred screen time and takes much longer. The cut is instantaneous and, if it is in the right place, completely unnoticeable.

It has been held that a mix is 'smoother' on music, but this is another fallacy. The mix is always obtrusive and, therefore never smooth; whereas the cut, if it obeys the rules and is correctly timed, is virtually invisible and, therefore, completely smooth.

There is, apparently, another strange school of thought which seems to hold that, in some way, the cut is impious and the mix is reverent and that, as soon as cameras are brought into a place of worship, the cut must be eschewed and the mix must be used as the only means of getting from one camera to another, with the result that a material proportion of the service is blurred.

Newcomers to the profession can be assured that neither the cut nor the mix is the moral superior of the other. Each has its place; each has an entirely different meaning.

The misuse of the dissolve or mix confuses the viewer and destroys its significance and usefulness. It is the sign of bad and unprofessional workmanship.

8b. Although the essence of television is the close-up, do not overlook the value of the long-shot.

Although the impact of the close-up is greater than any other type of shot, too much close-up can be worrying and bewildering. The eye needs a change and an occasional rest from close-up and the intervening long-shot makes the ensuing close-ups, more effective and welcome.

Furthermore, without long-shots, the viewer gets no sense of geography or of the physical relationship of the persons concerned. It is a good thing, now and again, to remind the audience of these things with a long-shot. Exits, entrances and major moves are good motivations for cutting to the long-shot.

8c. As soon as possible after going to a new scene, give a long-shot of it.

This tells the audience where they are and what the place looks like. It gives them geography.

8d. As soon as possible after the entrance of a new character of any importance or after the re-entrance of a character who has not been seen for some time, give that character a close-up.

When a new character comes on, the audience, instinctively, wants to see what he looks like. Satisfy that instinct.

In the case of the re-entrance of a character who has not been seen for some time, the close-up is useful as a reminder to the audience as to what he looks like and as a safety precaution in case he was not recognized in long-shot.

8e. Try to avoid cutting from a very long shot of someone to a very close shot of him.

The effect of this is ugly and the close shot seems to jump at one. Either have an intervening mid-shot or bring the subject into the foreground of the long shot before cutting to the close shot, unless a deliberate shock effect is wanted.

Never cut to a close shot of someone unless he is recognizable and readily identifiable from the preceding long shot.

9. Avoid panning or tracking with a lens giving a horizontal angle of acceptance of more than forty degrees.

The very wide angle lenses tend to bend and curve both horizontal and vertical straight lines. This is hardly noticeable when the camera is stationary but is immediately noticeable as soon as it moves, as the degree of curvature varies in different parts of the picture with the result that the straight lines appear to bend and twist. A very ugly effect.

10. Do not track in or out with a lens giving a horizontal angle of acceptance of less than twenty degrees.

The reasons for this are: (1) Because the angle is narrow, a track will make very little difference unless it is very long. (2) Because the distance from the camera to the subject is great, any unsteadiness in the camera will be greatly magnified in the picture. Imagine a hand holding a long fishing rod. A tiny movement of the hand will result in a huge movement at the end of the rod. So it is with the narrow angle lens, the smallest jog of the camera due to an unevenness of the floor will result in a great leap of the picture.

It is true that on a lens giving a horizontal angle of acceptance of fifteen degrees, given a perfect floor and a very good cameraman, a tiny creep in might be attempted, but it is better avoided.

11. Do not over-use a zoom lens as a substitute for tracking. Regard it, rather, as an infinitely variable lens turret—something

to be operated when it is not on the air—except for deliberate effect.
If the zoom lens is used, on the air, it has a very strange and unnatural effect. It causes the horizon and the middle distance to approach or recede with the same velocity as the foreground which is not the case in tracking or with the human eye in travel.

It is true that at some sporting or other unrehearsed event the exigencies of the circumstances may make it necessary to use it in this way but the effect is always unpleasant and should, if humanly possible, be avoided.

A very slow, gentle 'creep in' with a zoom lens would, probably, not be noticed but fast movement on it is an abomination unless it is specially used to achieve some rare and weird and unnatural, dramatic effect like a zoom into the close-up of a screaming face. Except for this sort of shock treatment, the rule is '*keep it still*'.

12a When it is necessary to take a comparatively close shot of someone from a long distance away, by using a narrow angle lens, the camera should be as nearly on a level with the subject as possible.
A very noticeable feature of narrow angle lenses is their foreshortening effect. They bring the background up close to the foreground. A close shot of a man standing ten feet from a wall will, on a narrow angle lens, make him appear to be standing right up against it.

Similarly, a full length shot of a tall man taken from high up on a narrow angle lens will make him look like a dwarf with very short legs. This fault sometimes occurs when shooting in theatres with the camera in the dress circle or gallery and at such things as cricket matches when the camera is often on the roof of the pavilion.

12b. When using a very wide angle lens, one with an angle of acceptance of more than forty degrees, if a natural perspective is desired, the camera should be as nearly horizontal as possible.
The very wide angle lenses distort perspective. This distortion

is not noticeable so long as the camera is horizontal. The moment it is tilted to any degree the distortion is very noticeable. If it is necessary to take, say, a long shot of the stage of a theatre it is better to take it from the dress circle rather than the gallery and better still to take it from the stalls rather than the dress circle.

Relate this rule to rule 9. *With very wide angle lenses the camera should be both horizontal and stationary.*

Note. While this distortion of the wide angle lenses can be objectionable it may, in certain special and rare circumstances, be very useful. A very wide angle lens on a sharply tilted camera, especially one tilted upwards from a low set-up, can, by very virtue of its distortion, give a powerful and dramatic, if bizarre, effect.

13. For all shots, except individual close-ups and those in which the objects of interest are in the same plane, use a lens with a horizontal angle of acceptance of more than twenty degrees.
The narrower angle lenses distort, diminish the perspective and, therefore, the relative sizes of the objects in depth.

The, possibly, permissible exception to this rule is when shooting over-shoulder shots—a close-up of A over B's shoulder. Here a lens giving an angle of acceptance of fifteen degrees will tend to enlarge, and emphasize A's face, which is important in relation to the back of B's head, which is unimportant. This, though it is optically wrong, is often dramatically effective. However, to go beyond this to an even narrower angle lens would be noticeable and look rather absurd.

Remember, also, that the relative sizes of A and B are governed by the distance that they are apart—the greater the distance, the greater the difference—therefore, though a fifteen degree lens might well be acceptable when A and B are five feet apart it would be much less acceptable when they are two feet apart.

Here is an impression of the same shot taken on three different lenses with A and B about three to four feet apart.

Fig. 1

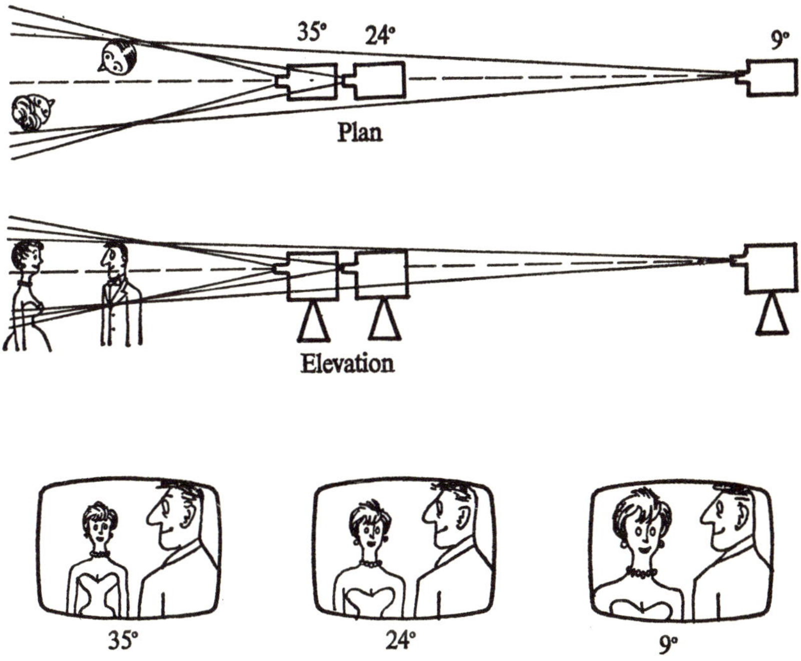

14. Allow time for lens changes.

Some television cameras are equipped with rotating turrets capable of mounting four lenses of different angles (a nine, sixteen, twenty-four and thirty-five degrees is a normal combination) any of which can be brought into use by the cameraman revolving the turret. This process is called 'swinging lenses' and takes several seconds. It can only be done when the camera is 'off the air' long enough for the lens-turret to be 'swung' from one shot to another and for the camera to be newly lined up and focused on its fresh subject. Even with a zoom lens it is necessary to give the cameraman time to change the angle and set up the shot before coming to it.

Zoom lenses, of course, make an equivalent change of angle very much more quickly. A modern zoom has a continuous range from five degrees to fifty degrees and can be pre-set for any point in between. The cameraman presses a button and the new angle is available almost immediately. This also has to be done when the camera is not 'taking' (except for a special effect) or there would be an ugly jump in the picture.

It needs to be borne in mind that zooms, however useful, have disadvantages in operation. They were originally designed for static cameras and have a near focus point of at least three feet. They are therefore cumbersome in developing shots (i.e. when the camera is moved about while 'on the air'). A normal fairly wide angle lens can move around its subject in a much smaller arc and thus calls for less movement of the camera and camera mounting.

15. On close shots of people, avoid both too little and too much head-room. Remember that the head of the subject should never touch the top of the frame and the chin of the subject should never touch the bottom of the frame, unless the shot is so close that the upper and lower edges of the frame cut both the chin and the forehead.

Wrong

Wrong

Right

Right

15a. Make certain that all the cameramen in a team have the same idea of head-room allowance and stick to it.

It can be very worrying to the viewer if cameramen on the same production have different conceptions of what is correct. Heads will appear to bob up and down in the frame on cuts and the effect will be ragged and untidy.

15b. In an individual close shot, if the subject is looking to the right, compose the picture with him slightly off-centre to the left. If he is looking to the left, compose the picture with him slightly off centre to the right.

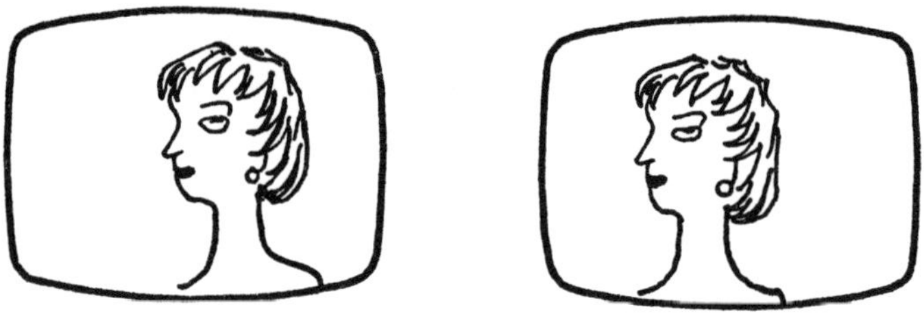

Prefer this to this

Quite apart from the fact that it is more pleasing to the eye not to have the subject's face squashed up against the frame, if this rule is not observed when cutting between a pair of close-ups, the two subjects may appear to be standing back to back.

15c. Avoid shots in which objects appear to grow out of the top of a person's head, because they are on the same line as the subject and the camera.

Frame 1

Frame 2

It will be agreed that Frame 1 looks somewhat ridiculous. If, however, we move the camera slightly to the right or the artist slightly to the left, we get Frame 2 which is a good deal better.

15d. Avoid shots that are wider than they need be to contain the action and, even more rigorously, avoid shots that are too close to contain the action.

For the sake of clarity it is obviously desirable for all shots to be as close as is reasonably possible. On the other hand, it is irritating to the viewer if the camera is too close and action that should be seen takes place just outside the frame. It is annoying if someone is reading something and we can't see what he is reading, or pouring out drinks that we can't see. It is irritating and ugly if hands making expressive gestures do so outside the confines of the frame.

A common fault is the close-up which is too close and does not give the head room for natural movement in animation. This causes the cameraman to be continually panning to keep the head in frame and the effect is restless and distracting.

The ideal length of a shot is one that will just contain the essential action.

16. In general, avoid shots in which people, seen full face or nearly full face, are cut vertically by the edge of the frame.

Wrong

Right

This is a particularly ugly and irritating effect which is caused by the camera trying to be closer than it should be. It always looks rough and untidy.

The exception to this rule is when shooting a crowd. With a natural crowd it is unavoidable and not objectionable. With a simulated crowd in the studio, composed of comparatively few people, it is a useful device which makes the crowd appear to continue and be bigger than it is.

It is when shooting a group of, say, five or six people that it looks particularly bad if those on the extreme right and left are cut vertically by the frame. In this type of shot it is better, if the camera must err, to be too wide than too tight.

16a. In group shots, compose in depth and avoid the straight line.

You will agree that this

is a good deal better

than this

16b. Avoid grouping artists in such a way that someone who is not essential to a shot will be half seen in the background and half masked by someone in the foreground.

Avoid this

prefer this

or this

It looks very untidy and can be very annoying and distracting to the viewer to have someone lurking about, half seen, in the background of a shot. If he is essential to the shot or can contribute to it, arrange for him to be completely visible. If not, engineer him out of the shot altogether.

16c. Balance the objects of interest in the frame. Avoid getting them bunched up in the middle with nothing on either side, or, what is even worse, getting them at the extreme edges of the frame with nothing in the middle.

Prefer this

to this

or this

But, better still, because it is closer, prefer this

The 50/50 two-shot, as it is called, tends always to be too much in profile and facial expression is lost. Best of all, therefore, prefer a pair of matched and complementary over-shoulders like these.

16d. Provide foreground objects in very long shots.

A very long shot without anything in the foreground is apt to be dull and ugly and rather flat with a vast expanse of floor or ground at the bottom of the picture and all the interest in the distance at the top. Foreground objects, a shrub or a piece of furniture, hide the floor or ground, give interest to the bottom of the picture and depth to the composition.

Compare these two pictures, the one on the left with foreground objects, the one on the right without.

16e. When composing in depth, consider depth of focus.

The depth of focus is the distance between the nearest point to the lens which is in focus and the furthest point from it which can be kept in focus at the same time.

The depth of focus varies with:

A. The lens. The wider the angle of lens the greater the depth of focus.

B. The distance of the lens from the subject. The greater the distance the greater the depth of focus.

C. The intensity of light. The greater the intensity the greater the depth of focus. If the light is bright the lens can be stopped down, that is, the lens aperture decreased. This gives greater depth of focus. If the light is less bright and a larger aperture has to be used, the depth of focus is less.

If you are planning a shot in great depth, particularly a

fairly close shot or one on a narrow angle lens, get technical advice as to whether or not it can be held in focus.

Although sharp focus is generally desirable throughout the depth of the shot, it is not always so. In a close over-shoulder, for instance, provided that the person being favoured, that is the person facing the camera, is in sharp focus it will not matter if the person with the back of his head to the camera is slightly soft. If, however, they were both facing the camera, it would be most objectionable if one of them were out of focus.

In a close-up, it can be a definite advantage to have the background out of focus as this concentrates the attention of the viewer on the face of the subject and blurs what might be distractions in the background. Faces, however, should be in sharp focus.

Do not be misled by the pocket view-finder. This is in fixed, sharp focus throughout its entire range and can be no sort of guide.

16f. Let the lighting help your shots.

The distribution of light over a scene can do a great deal to increase or reduce the effectiveness of the shots. If the lighting is too flat the subjects of your picture will not stand out. If it is too localized important actions or expressions may not be seen.

In outside broadcasts or filming you will often have to accept the light as you find it, but in the studio you have control over your lighting effects and it is important to consult your lighting supervisor beforehand and to plan your shooting with him so as to get the results you want. For instance, if there is a rational source of light, such as a window, try to arrange your scene so that the key lighting from there will favour your main action.

Changes of lighting tone from scene to scene will give variety and heighten the style of a programme; cross-lighting on a set, such as from doors opening on to a corridor, will give added depth to a scene; high-lighting or silhouetting of chosen pieces of action will add to their dramatic effect. To achieve all this it is essential to make sure that the lighting supervisor and the scenic designer both fully understand in advance what you are trying to do.

Nowadays most television cameras (working in monochrome) can operate effectively at an aperture of f.5·6 and the studio will be lit for you at this level. This will give you a comfortable depth of focus, but if you require special deep-focus effects you must warn your lighting supervisor in advance. Colour cameras do not operate at f.5·6.

CAMERA WORK

17. Do not pan across a static scene merely to get the camera from one point of interest to another. Only pan with a moving person or object.

The reasons for this rule are:

(A) As soon as a camera moves independently in this way it draws attention to itself and the technique and distracts the mind of the audience from the subject and the content matter.

(B) Such a pan can seldom be motivated—see rule 8.

(C) To pan across a static scene is unpleasant to the beholder because it is something that the human eye does not do. Try it. Focus your eyes on an object on one side of the room, then turn your head and focus them on an object on the other side of the room. You will find that though you have seen the two objects, you will not have been conscious of having seen what has come in between. In fact the eye will not have panned, it will, virtually, have *cut* from one object to the next. On the other hand, the human eye is perfectly capable of panning when it is following a moving person or thing.

The 'whip pan', a movement whereby the camera sees one object, then pans very quickly and sees another, is, though a doubtful and ugly manœuvre, legitimate because it does what the eye does and blurs the intervening detail. It is like a cut with a smudge in the middle.

The very, very slow pan across a distant landscape is tolerable because the eye can do it provided that it moves slowly enough and the scene it is looking at is sufficiently far away. What it cannot manage is any sort of fast movement in an enclosed space. Therefore, never pan across a static scene in a studio set except in very exceptional circumstances (e.g. when moving from object to object in order to build up a particular effect or to provide backgrounds for a series of captions).

18. Avoid the fast pan.

This causes 'strobing', a streaking, juddering effect which is ugly. The same thing may be caused with roller captions that are moving too fast. (This is particularly important when a programme is being recorded on film, as a trailing effect is liable to occur, even on modern equipment, on objects moving at certain speeds).

19. Do not track *back* unless it is with a person or object that is moving towards the camera, or while panning with a group that is widening during the track, or when the track back is palpably motivated by the action or the dialogue.

This rule is, of course, an extension of rule 8. Never do anything without a reason that is clear to the audience. Never, for instance, track back from a two-shot in order to see the door because you know that, in a moment, someone is going to come in at it. The reason won't be clear to the audience till afterwards when it will be too late, and the track will have diverted the audience's attention from the matter of the two-shot. You should cut to a long shot, in this instance, on the 'click' of the door-catch, the 'click' being the motivation for the cut.

The reason that the track-back is particularly bad in television is that the definition of the picture is imperfect and the human eye is always asking to see more clearly, in other words, to get closer in. Whereas the track-in satisfies this instinct and is motivated by it, the track-back does the reverse and irritates the beholder.

The track-back with a person or object moving towards the camera is legitimate, because the camera is only tracking away from the background and is maintaining its same distance from the subject of interest. Furthermore, the movement of the person or object motivates, and gives a palpable reason for, the track. Similarly, the track-back while panning with a moving group that is widening is legitimate, because the very fact that it *is* widening motivates the track. If there were no track, part of the group would go out of shot. This really comes under the heading of a track-back that is motivated by the action.

Another example of a track-back motivated by action is that of, say, a medium close shot of a seated person. He rises, and it is necessary to pull back in order to contain his rise, the rise motivating the track.

It should be noted that, though this is a legitimate manœuvre, it is a difficult and slightly risky one on live television. It must be carefully rehearsed and timed. If the track-back is too early, it will not be motivated and will be a distraction. If it is too late, the rising person may, for a moment, put his head out of frame, which is untidy and unprofessional. To be right, the track must be absolutely on and with the rise. It would be safer to cut to a longer shot *on* the rise.

An example of the track-back motivated by dialogue: imagine an art historian giving a talk on painting. He shows us a close-up of a detail of the painting and discusses it. Then he says: 'Now let us stand back and look at the picture as a whole.' Here is a perfectly legitimate reason for the camera to pull back and do what our human eye has been asked to do. It will not irritate the spectator because he will want the camera to track back.

19a. Avoid restless camera work.

The camera in motion can be extremely effective but it is possible to have too much of a good thing. It can also be irritating and distracting. It can divert the attention of the audience away from the subject and towards itself and its own antics. This, however clever it may be, is bad camera work. Learn rule 8 and never move the camera unless there is a good reason for it, a reason that is obvious to the audience, a motivation that springs either from the action or the dialogue.

20. In no circumstances whatsoever cut between cameras which are both giving practically the same shot. Only cut between cameras that hold either:

(A) Totally different subjects, or, if the subjects are the same, that have

(B) A major and material difference in length (i.e. size of framing) between the shots, or

(C) A major and material difference in angle between the shots.

This is, perhaps, the most important and the most binding rule of them all. The reasons for it are:

(1) If both cameras are giving virtually the same shot there can be no possible reason for cutting, as the second camera will not be saying anything different from what the first has been saying. The cut will be unmotivated and meaningless and will only constitute a distraction from the subject.

(2) The more powerful reason against cutting between cameras giving very similar shots is that the effect is discordant and ludicrous. Such a cut does not look like a cut, it looks as if the people and objects in the picture have spasmodically jumped to a slightly different part of it.
To state the rule in terms of some instances you may

(A) Cut from, say, a close shot of X to a similar, matched, close shot of Y because the subjects are different. You must never cut from a close shot of X to another, similar close shot of X.

You may cut from this to this

Difference in subject

But never from this to this

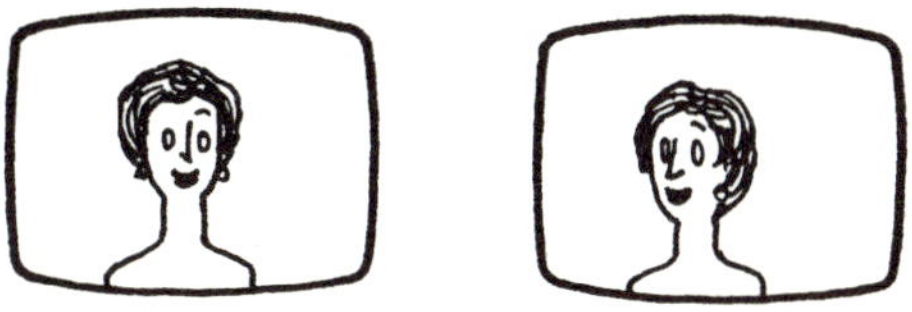

Similar

(B) You may cut from a tight three-shot of X, Y and Z to a medium long-shot of X, Y and Z because the length, the framing, is different. You must never cut from a three-shot of X, Y and Z to a similar three-shot of X, Y and Z.

You may cut from this to this

Difference in length

But never from this to this

Similar

(C) You may cut from a close shot of X over Y's shoulder to a close shot of Y over X's shoulder because, although the framing is the same there is a material change of angle. You must never cut from a two-shot of X and Y to a similar two-shot of X and Y. It would just make them appear to jump slightly.

You may cut from this to this

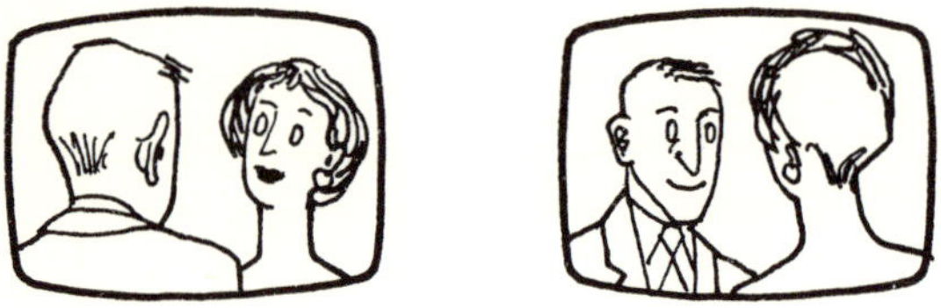

Difference in angle

But never from this to this

Similar

21. In cross-cutting sequences, prefer matched shots to unmatched shots, complementary angles to angles which are not complementary.

By 'cross-cutting sequences' is meant sequences in which one cuts back and forth many times between the same pair of shots, as for instance on static duologues. In such sequences if the shots are not matched, the cuts will be very noticeable and the effect will be jarring and unpleasing. If they are matched the cuts will be unobtrusive and the effect will be smooth.

If the angles are not complementary, the characters will not appear to be looking at or talking to one another. If they are complementary, then they will appear to be looking at one another.

This page is RIGHT

Fig. 2

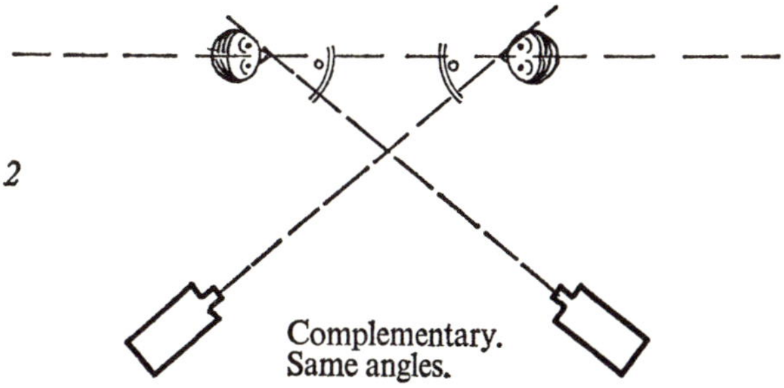

Complementary.
Same angles.

Fig. 3

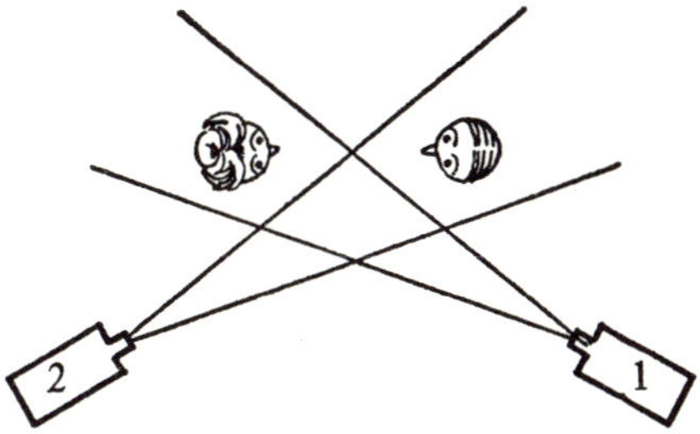

Complementary.
Same angles.
Matched.
Same size.

1 

2 

Fig. 4

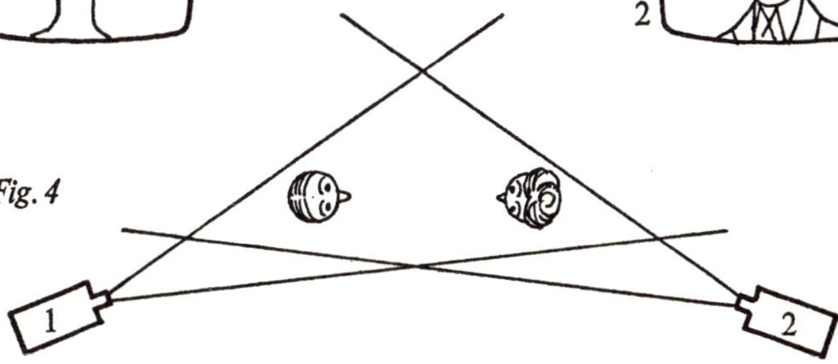

Matched over shoulder shots.
Same lens, therefore same perspective.

1 

2 

This page is WRONG

Fig. 5

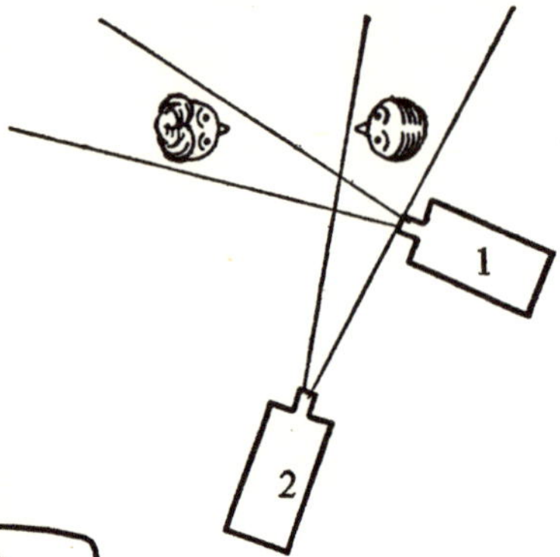

1 

Not complementary.
Same size but different angle.

 2

Fig. 6

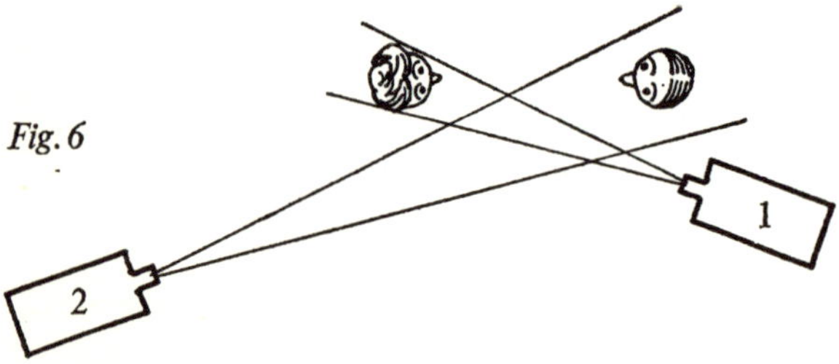

1 

Unmatched.
Same angle but different size.

 2

Fig. 7

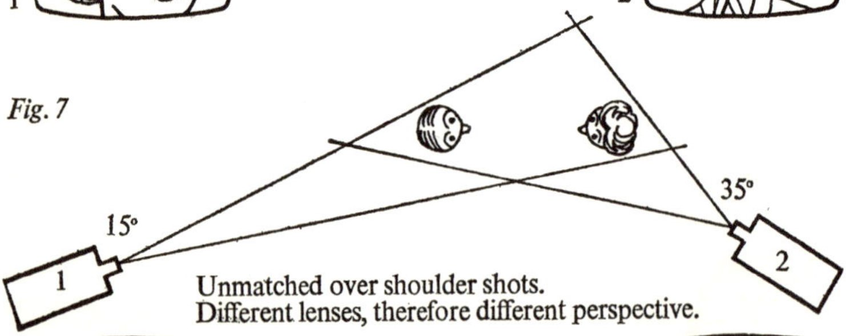

Unmatched over shoulder shots.
Different lenses, therefore different perspective.

1 

 2

Matched shots are those in which the size of the artist, relative to the frame, is the same in both. They should also, if possible, be matched as regards tone of background and faces.

In the case of over-shoulder shots, the perspective should also be the same. Therefore use for both, the same lens at the same distance. (See Figs. 1, 3 and 4.)

When it is necessary to change the size of matched shots (e.g. to move in from mid-shots to close-ups) the the change should be made on *both* cameras at a suitable dramatic moment. An exception to these rules can be made when a 'personality' is being interviewed. In these circumstances it is often rewarding to explore the 'victim's' face with a close-up lens while keeping the interogator unobtrusively at a distance.

Complementary angles are those in which the eye-line of the artist is the same number of degrees 'off lens' in both cases though, of course, in opposite directions.

To put it another way, imagine a straight line drawn between two people who are looking at and talking to each other. With complementary angles, each camera will be at the same angle to this line. Only if it is will the people appear to be looking at each other.

One has heard it said by beginners that a pair of matched over-shoulders such as these are not possible in television because the cameras would be in each other's shots. This, of course, is not the case as both the cameras *must* be on the same side of the imaginary line between the two subjects. (See Rule 22.) Examine Fig. 4, which shows the correct disposition of cameras for this type of thing. They are perfectly clear of each other's shots.

22. At all costs, avoid so disposing cameras that the subject will be looking out left of frame in one shot and out of right frame in the next and vice versa.

This is one of the worst crimes that a television director can commit. It has a most disquieting and ludicrous effect. It may not, on the face of it, seem possible, but it *is* quite possible. Look at the diagram Fig. 8.

Fig. 8

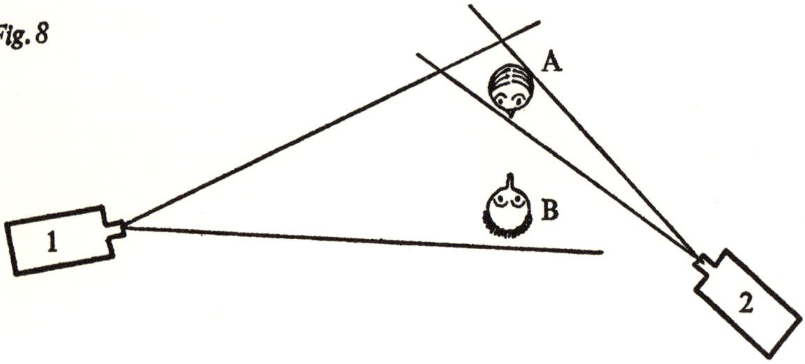

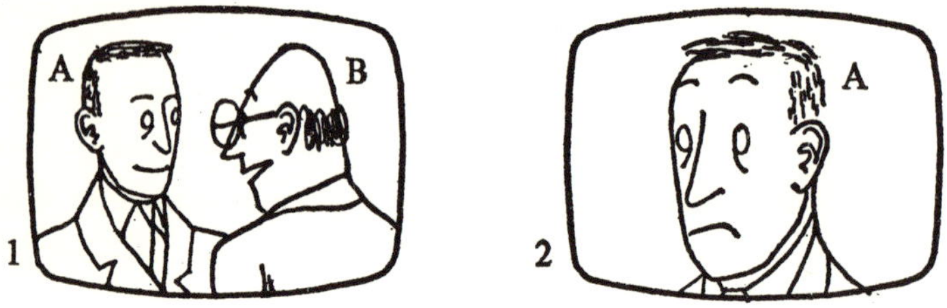

A and B are two persons having a conversation. Camera 1, on a wide angle lens, holds both of them in a medium shot favouring A. Camera 2, on a narrow angle lens, holds A in an individual close shot. Were we to cut from camera 1 to camera 2, A would be looking out right on camera 1 and out left on camera 2. He would appear, magically, to have turned round on the cut.

The rule, here, is to draw an imaginary line between A and B and make sure that *both* cameras are the *same side* of this line.

You will see that, in the next diagram (Fig. 9), both cameras are the same side of the imaginary line between A and B and that, therefore, A is looking from left to right in both shots.

Fig. 9

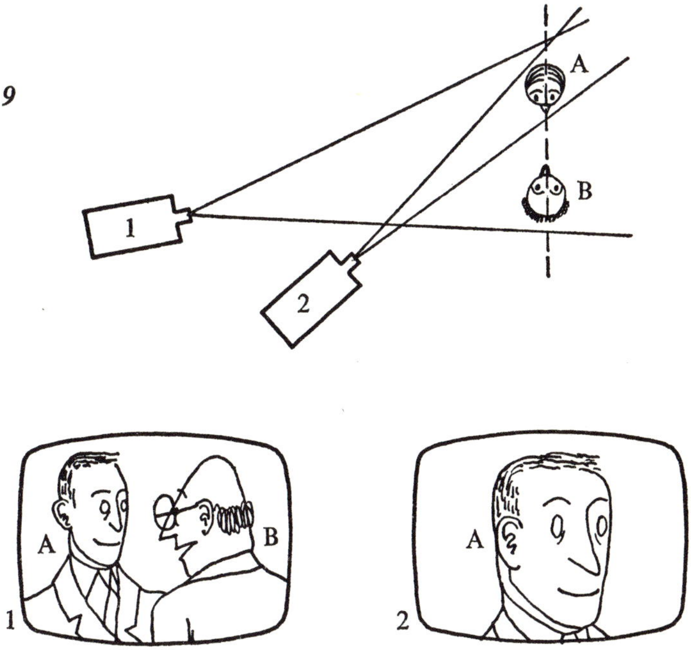

To take another common example. Imagine a pianist in a concert hall, playing the piano with two cameras on him, Camera 1 giving a long shot and Camera 2A giving a close-up, disposed as in Fig. 10, on either side of him.

Here, again, because we have 'crossed the line', the pianist is looking from left to right on camera 1 and from right to left on camera 2A and would appear to jump round through 180 degrees if we were to cut from one camera to the other. Were we to draw the imaginary line from the pianist to his piano and keep both cameras (1 and 2B) on the same side of the line, he would be looking the same way in both.

It is true that when shooting from, say, a public concert hall, a director may not always be able to put his cameras exactly

Fig. 10

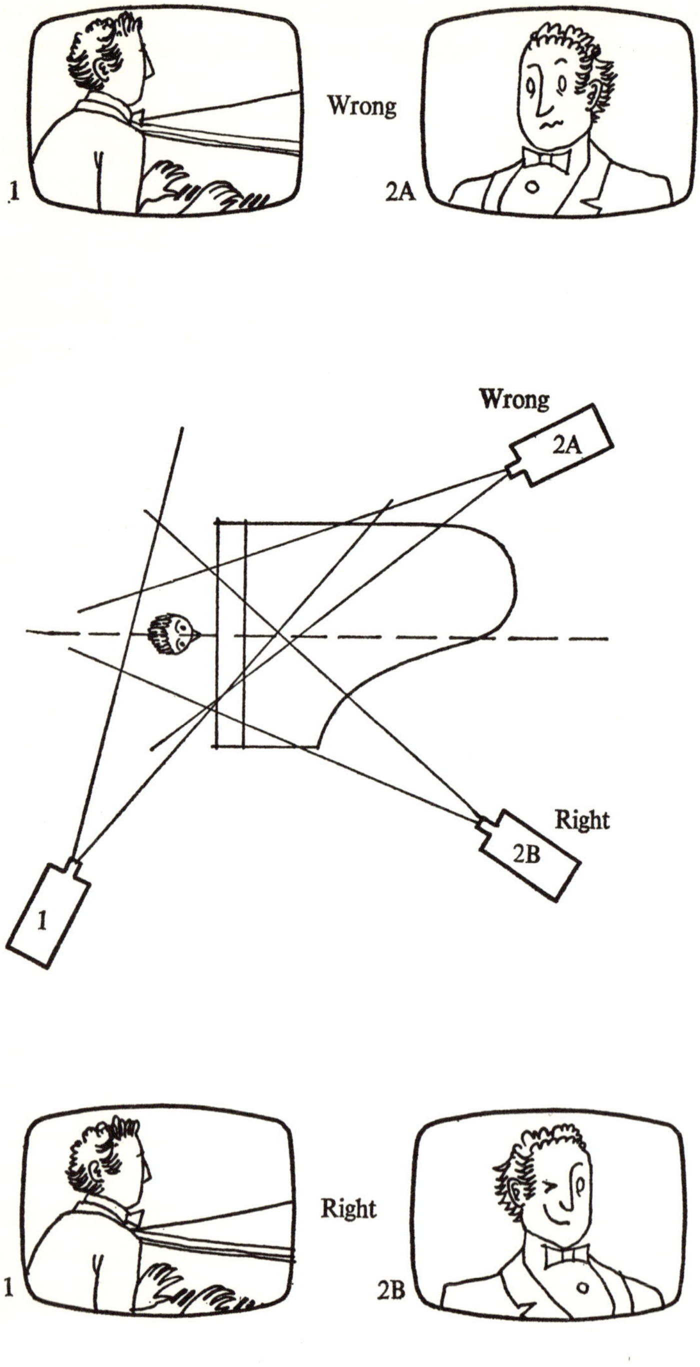

where he wants them. He may have to put them where he can, but that is still no excuse for cutting direct from camera 1 to camera 2A. The director will certainly have, at least, a third camera. If, between these shots, he were to cut away for a second to one of the audience, or, assuming the pianist is playing a piano concerto, to a shot of the conductor, the turn would probably not be noticed.

22a. When cutting between either end of a telephone conversation, arrange for one speaker to be looking out right of frame and the other speaker to be looking out left. In no circumstances should they both be looking in the same direction.

Though, in real life, the speakers might well both be looking in the same direction, if they are allowed to do this on film or in television the effect on the audience is most strange. They don't appear to be speaking to one another, but to some unseen third party. Subconsciously we expect people who are talking to one another to be looking at one another. If this subconscious urge is not satisfied, we disturb the audience and create a distraction.

You can well see that if we cut between this pair of shots

the two people will not appear to be speaking to each other, whereas, if we reverse one of them, they will. Thus:

22b. If several actors are looking off-screen at an object unseen by the audience, make sure that they all look in exactly the same direction—all have the same eye-line.

If, in a play, for instance, you have a group of actors looking off-screen at an object—the top of a high building say—it is essential that they all look in exactly the same direction, otherwise the effect is false and artificial. Place, out of shot, a small flag or disc or any readily recognizable object and tell them to look at that. The smallest deviation of eye-line is detectable but can be avoided by this means.

If the off-screen object they are supposed to be looking at is a moving one then the studio marker should move too. A travelling car can be marked by a man walking with a flag. If the actors in picture are watching Mr. Solness climbing his tower, then contrive for the marker to be hauled up by a rope and pulley so that all eyes look at the same point and tilt up at the same speed.

23. At all costs, avoid so disposing cameras that a person or thing will be travelling from left to right in one picture and from right to left in the next and vice versa—that they will change direction in mid travel.

The effect here is, if possible, even more absurd than the effect of breaking rule 22. The principle for avoiding it is the same, only in this case, all the cameras must be kept the same side of the line of travel.

Imagine a race-course with a group of horses galloping round it and the cameras disposed as in Fig. 11 (opposite).

You will see that, if the horses are travelling in the direction of the arrow, on camera 1 they will appear to be going from right to left and on camera 2 they will appear to be going in the opposite direction, from left to right. If

Fig. 11

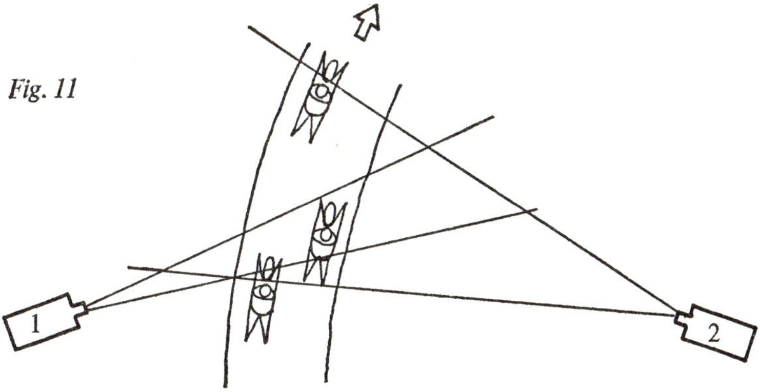

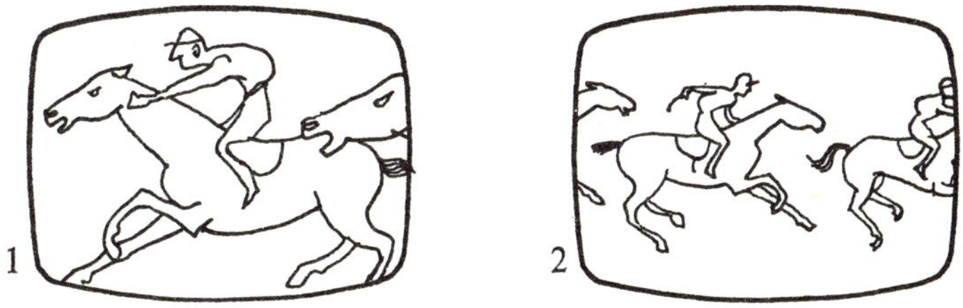

you cut between these cameras the effect will be grotesque.

Now try putting both cameras on the inside of the curve of the racecourse, both on the same side of the line of travel. (Fig. 12.)
Here, the horses are all going in the same direction on both cameras which tends to make for sanity.

These confusions of direction are slightly less likely to happen in a studio production with conventional, three-sided sets (though they can happen even here) but they may easily occur in outside events if the cameras are not properly disposed.

At a football match, a cricket match, a tennis tournament, the rule is: draw an imaginary line down the centre of the football field from goal to goal, down the cricket pitch from wicket to wicket, down the tennis court at right angles to the

Fig. 12

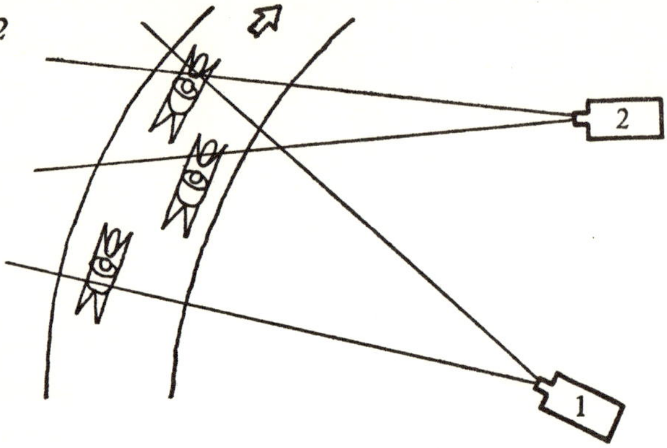

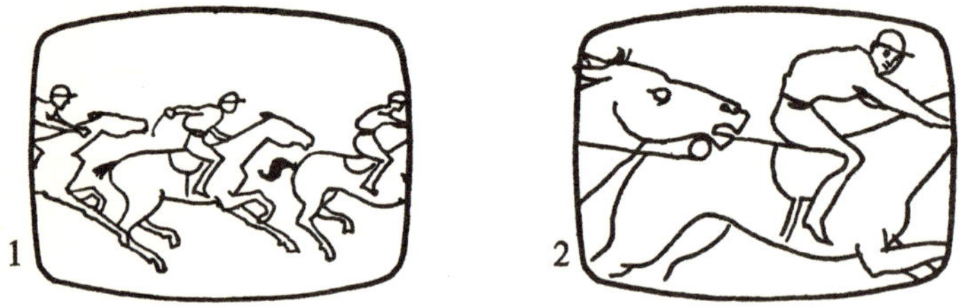

net and keep all your cameras the same side of this line. It does not matter how close they get to the line as long as they don't actually cross it, but the moment they *do* cross it you will be in trouble and the audience will lose all sense of direction.

Note: The foregoing remarks concerning programmes outside the studio, relate to what *should* be done. In actual practice, cameras may have to be put where they *can* be put and it is better to see the race or the game somehow—even if it is technically wrong—than not at all. However, this is a code, a manual of instruction, and it must concern itself, primarily, with correct method and state that correct method. It cannot possibly hope to cover the exigencies of every conceivable circumstance.

24. Avoid making a major point of interest jump from one side

of the frame to the other on a cut. If the point of interest is, say, on the left of the frame in one shot, it should be on the left of the frame in the ensuing shot.
This is an easy rule to break, an easy trap to fall into if one does not watch it very carefully and the effect is very disquieting and incongruous.

Look at the example below (Fig. 13). On camera 1, B is on the left of the frame and on camera 2, she is on the right, therefore, on the cut, she would appear to jump from one side to the other.

Fig. 13

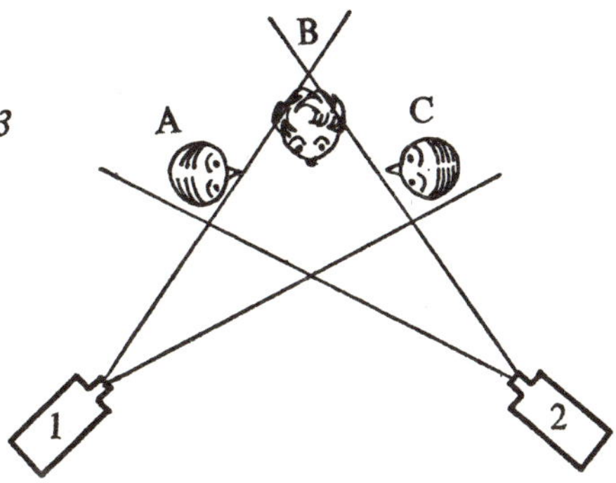

Still on Fig. 13, cover camera 1's frame with a couple of fingers and look at camera 2's frame. Now move your fingers quickly so as to cover camera 2's frame and expose camera 1's. Do this backwards and forwards a few times. You will get an impression of cutting from one to the other and you will see how B jumps from side to side.

The foregoing example is, of course, an extreme one because the two pictures are similar, and B is completely on the right of one and completely on the left of the other.

Fig. 14

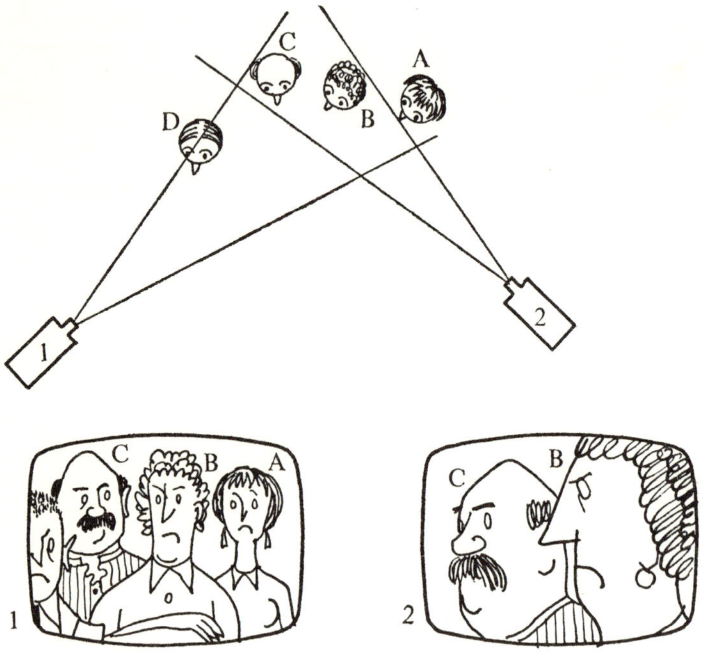

Examine this example (Fig. 14).
Here B is completely on the right of camera 2's picture but roughly in the centre of camera 1's. The jump is much less great. Furthermore the two pictures are quite different and the small jump would never be noticed.

This rule not only applies to a person or a major point of interest, it can also apply to an *obtrusive* object of relatively minor interest.

Fig. 15

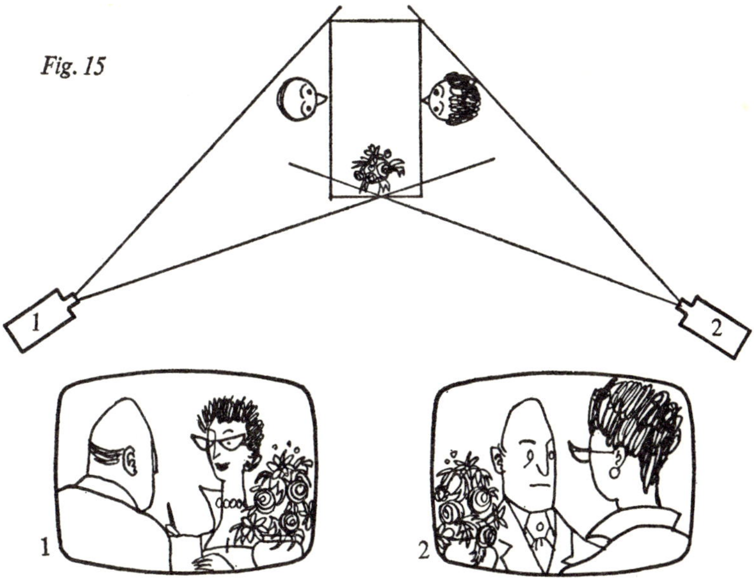

Imagine two people having a conversation, sitting either side of a desk and talking to one another across it. Imagine them covered by a pair of matched, complementary over-shoulders as in the diagram Fig. 15. Imagine an obtrusive vase of flowers on the end of the desk in both shots. In one camera it will be on the right of the picture, in the other camera it will be on the left. On each cut, the flowers will fly from side to side of the frame, which might well constitute a ludicrous distraction.

A small and relatively unnoticed object doing this would not matter but something like a vase of flowers, a bust or a lamp which rivalled the subjects in obtrusiveness, would matter very much indeed.

24a. On close-ups, keep the shot clear of distracting objects and parts of objects.

In television studios this is often difficult to do but it *can* be done if special care is taken. It can be infuriating if, on a close-up of A, one sees, at the edge of the frame, the coat or the sleeve or the stomach of B. It spoils the composition and frequently distracts the viewer's mind from what A is saying. This is particularly the case if only part of the sleeve or the coat or the stomach is visible and the viewer is trying to identify what it is. Keep the frame clear of anything that might take the viewer's mind off A.

24b. In over-shoulder shots, never let the head of one person overlap or mask the face of the other.
In this type of shot the actor or performer must not be allowed to lean too much to the right or left, as the case may be, or he will obviously mask or partially mask the person to whom he is speaking.

24c. Keep the angle of the camera as close as possible to the eye-line of the subject.
The more the subject is in profile to the camera, the more facial expression is lost. In general, where facial expression is important, prefer the full-face or nearly full-face shot to the profile shot.

24d. Never let a performer look straight into the lens of the camera unless it is necessary to give the impression that he is speaking directly to the viewer personally.
An actor in a play who is supposed to be speaking to another character should never in any circumstances look into the lens. If he does, realism will be lost as he will then appear to be speaking to the viewer. On the other hand, somebody giving, say, a lecture *should* look straight into the lens as he is addressing the viewer personally. Relate this to 24c. In a play, where facial expression is important, keep the camera as close on the eye-line of the actor as possible but never let the actor look straight into the lens. Provided that the actor does look just off the lens,

even if only by the tiniest fraction, he will not appear to be speaking directly to the viewer.

In a factual programme, when a subject is speaking and looking into a camera, it is disconcerting to change cameras. You are adding an unfair burden to speakers unused to television, and therefore unskilled at adjusting their eye-line, and the results can be unfortunate.

25. When planning the action, always bear in mind the problems of lighting.

Lighting for television is full of problems. It is very much more difficult than lighting for film where only one shot at a time is taken from one camera as opposed to television's multiplicity of shots from three, four or six cameras.

If you want good lighting, consider the problems of the lighting supervisor, consult him throughout all stages of the production and give him shots that he *can* light. (See 16f above).

Do not, for instance, get people close up against the walls of a set and then expect good lighting. In television, as in film, back lighting is an essential. This becomes impossible when a person is hard up against an opaque object like a wall.

As a good general rule, plan your action, whenever possible, well out in the open, as far from the walls of the set as you can manage.

If any part of the set overhangs the floor, avoid action that is underneath it, if possible. It will be very difficult to light. If it is essential to have a roofing piece, keep the action away from it. It will be practically impossible to light people who are beneath it.

Try to avoid having very small, enclosed, sets. If, for some reason, this is unavoidable, keep the flats, the walls of the set, as low as possible. This enables the lighting engineer to get his lamps low and prevents the light hitting the subjects too steeply.

26. Don't forget the sound crew.

However exacting the demands of cameras and lighting, it is

equally important that the microphones should be able to pick up the sounds you want heard. In fact, the same rules apply. Words spoken under ceiling-pieces, against back walls or in deep narrow sets require microphone booms stretched out into the sets to an extent which makes it very difficult to keep them, or their shadows, out of the shots. Valuable time is often wasted in the studio because these difficulties have not been foreseen, and the outcome is frequently a partial withdrawal of the microphones and consequent deterioration of the sound. Sometimes, in these cases, viewers at home cannot hear the speech clearly.

Words spoken away from microphones, or conversation between two people on the move or widely separated, are obviously difficult to pick up. It helps if speakers and actors are made aware of their microphone's position and reminded to 'favour' it—i.e. not to speak *at* it in an exaggerated way, but to avoid speaking away from it. Above all, it is essential to consult the sound supervisor before and during the production, and to make sure that his crew can achieve good quality sound coverage without undue physical difficulty.

In outside broadcasts and filming particular care has to be taken to ensure that satisfactory sound reproduction can be obtained on locations which may have odd acoustics or noisy settings or be exposed to wind and weather. Advance visits to 'recce' shooting locations are as important for the sound as for the pictures.

27. Remember that there is perspective in sound as well as vision. These should always be matched.

Close-up sound, sound close to the microphone, has quite a different quality from long-shot sound which is distant from the microphone, therefore avoid having a shot in close-up with the microphone far away. It is pretty well impossible to do the reverse (except with lanyard microphones—see 30c below) as the microphone would be in shot; but close-up vision must always be matched with close sound.

If, by means of a narrow angle lens, you wish to take a close shot, deep in the set where the boom microphone cannot reach, arrange for a concealed microphone, which will give the right perspective, to be in position for the shot. Microphones can be concealed in lamps, chandeliers, vases of flowers, all sorts of things.

Never let different sounds swamp each other. If you want a mixture of speech and music, or of music and natural effects (such as the noise of a train), make up your mind which is to be heard more clearly and which is to be the background. Avoid, if possible, playing speech, music and effects all at once.

27a. Always relate the sound to the vision.

In relating pictures and sound, particularly in a programme with illustrated narration, make sure that the two help each other and do not conflict. A picture will make an immediate impression on a viewer, whereas a spoken comment will only make its impact when it has been completed and understood. On the other hand, a picture will mean more when the viewer has been prepared for it. Thus it is sensible to make the sound lead up to a new image and then leave the picture to make its impact undisturbed. After that, cue the speaker to make his comment and to prepare us for the next visually important moment.

28. Don't produce for the monitor in the control gallery; produce for the domestic receiver.

Always remember that the control room monitor is giving a better picture than will be seen on any ordinary, commercial, domestic receiver.

If it is essential to see facial expression, for instance, remember that, though you may be able to see it in a medium long shot on the monitor, it will probably need a considerably closer shot on the domestic set.

Remember, too, that on many domestic sets, the picture is expanded and there is cut-off at the edges. Do not, therefore, have essential action taking place on the very edge of the screen;

it may never be seen, particularly if the programme is being recorded on film. (See 31c below). Compose and frame your pictures with this point in mind.

Also bear in mind that if you are doing a dark scene—moonlight for instance—you are viewing the picture in a darkened control gallery whereas the viewer at home may not be watching his set in the same lighting conditions, and also his set may not be as finely adjusted in terms of brightness and contrast.

It is wise to write this type of dark scene out of the script if it is humanly possible. Sometimes, for plot reasons, this cannot be done. It is often impossible to avoid the night scene in some types of outside broadcast. It is quite impossible to avoid it in showing film that is not shot specially for television. If the scene can be got rid of, do so or keep it as short as you can. Often it is possible to change the time or locale of a scene so that it can be brightly lit. Sometimes it is practicable on an exterior to provide lanterns or brightly lit windows to the same end.

29. Make the scenery fit the production, not the production the scenery.

When working in a real setting, as in an outside broadcast, the shooting plan is dictated by the shape of the locale. In the studio it is limited by the space, time and money available for the sets. The wise studio producer will not waste his ration of settings on items which merely reproduce life but do not show to advantage in the pictures which the cameras will take, or in elaborations which simply get in the way.

In planning a production it is essential to keep the designer closely in touch with your plans and intentions. In fact, you should both work as a team with the technical supervisors responsible for cameras, lighting and sound. Consider not only the shape of your sets but their style and atmosphere (light and airy or dark and gloomy, and so on) and above all their adaptability to the individual shots you hope to take. Remember that in television what is not seen does not exist. Remember too that with a cluttered or fussy set the details are liable either to be lost or to be embarrassingly in the way on the small television screen: the strongest characters are often the simplest.

The conventional 'box' set—the building of a complete room with the fourth wall taken away so that cameras and microphone booms can poke their noses into it—is still widely used, but it is a legacy from the theatre and not always the most useful setting for television. As camera work has grown more fluid and enterprising, and as discontinuous recording has made it possible to pause between scenes to make changes in the studio (see 31 below), it has increasingly become the director's business to conceive his production as a series of independent visual set-ups, and not to anchor it within the rigid shapes of stage scenery.

Thus in a play it may be more effective to reduce a solid room to fragments of backgrounds against which the action can be played with space for the cameras close on each side of the actors, rather than to keep it as a solid three-sided structure where the cameras are held back at a respectful distance by unnecessary flats and furniture.

Sometimes sets can be dispensed with altogether and the action played against impressionistic wisps of scenery set in an open studio. However, the rules of grammar should be observed even more carefully in this sort of production than in programmes in conventional settings; for as other landmarks disappear the viewer needs particularly clear and lucid shooting, with understandable continuity between the shots, or he is left in a state of some perplexity.

USE OF STUDIO EQUIPMENT

30. Television studios vary considerably in their size and facilities, and the producer or director to-day is faced with a range of technical equipment which at first can be bewildering. It is not the purpose of this book to deal in details of engineering, but the following references may be useful in helping the newcomer to find his way.

30a. Cameras.
Most studio cameras are now fairly standardized and can be used as outlined in earlier chapters. Remember that—unlike the team in the producer's gallery—cameramen do not have scripts. They work from eye and ear or, at best, from cards which give a list of their own particular shots, and they need to have not only the initial set-ups but any alterations or developments explained to them.

Light, semi-portable cameras are sometimes available for special purposes, particularly on outside broadcasts; but the normal camera is still a fairly weighty piece of equipment and allowance needs to be made for its size and comparatively ponderous movement—and for its cable.

30b. Camera mountings.
The standard mounting for a television camera is a pedestal, which has three wheels that act like castors, and a ring around it by which its sole operator can push, pull, raise or lower it. It needs an operating space wider than its wheelbase (the cameraman has to stand somewhere), it cannot take the camera very high or very low, and it needs to work on a level surface (even so, it cannot be relied upon for a long steady track backwards); otherwise it can go almost anywhere and do practically anything.

Fig. 16

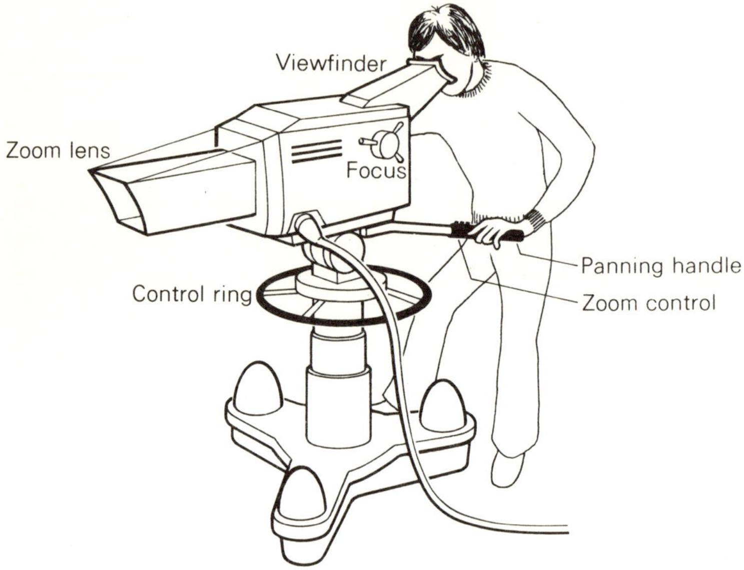

(In order to show the focus handle, it has been purposely drawn on the wrong side of the camera)

More ambitiously, cameras may be mounted on cranes or on four-wheeled motorized 'dollies'. These vehicles are much larger and heavier and give far greater vertical range and steadier tracking. For spectacular productions, or for programmes which need a good deal of steady 'dollying' in and out or a mobile camera at a high angle, these machines are extremely useful; but considerable space must be kept free for their manœuvre, and most of the motorized vehicles are limited in their angles of turn. New mountings are continually being developed and these give increasing freedom and variety of movement.

The four-wheeled 'camera-dolly' was inherited from the film industry, where tracking lines could be laid down individually for each shot, and the pedestal mounting used to be regarded as a poor relation, to which the junior cameraman was usually relegated. It is now generally appreciated that the despised pedestal (of improved design) is the most handy and versatile of studio camera mountings. (See Fig. 16.) The wise director, however, is not bound by prejudice or fashion but chooses the best mounting for the particular work he wants a camera to do, in consultation with his technical crew.

30c. Microphone booms.

Mobile microphone-boom platforms are now available with arms with a reach of twenty feet, but they need space in which to operate, which is rarely allowed them in the planning stage. Remember that booms do not go round corners, that the operator has to be able to *see* the microphone, and that no microphone platform (or camera mounting) moves entirely silently. Avoid, therefore, large migrations of booms (or cameras) from one set to another during quiet scenes.

'Lanyard' microphones, slung round the speaker's neck, are often used with success by professional commentators. They are, of course, visible. Also available are radio microphones, which are much less obtrusive. The sound quality is not as consistently good as that obtained by means of a boom—and there is a problem caused by rustling clothes. However, in some situations, these disadvantages are outweighed by the additional flexibility of movement permitted to the performer.

30d. Back (and front) projection.

The projection of still or moving pictures from the front (or, more frequently, from the rear) on to a large screen in the studio is a useful adjunct to scenery and can give a great variety of backgrounds. However, the screen, the projector and the long beam of light thrown on one from the other take up a great deal of studio space and are not moveable during shooting. The throw of the beam can be folded by the use of mirrors and in a discontinuous recording some scenes can be set and played and then cleared, but generally speaking the heavy apparatus of front or back projection is only worth house-room if it is to play a large and vital part in the production.

One cardinal rule when shooting against a projected background is that the taking camera must always be as nearly as possible at right angles to the background screen. If shot from an angle not only does the projected image become distorted, but the light coming through (or off) the screen into the camera lens is diminished and the picture simply becomes a grey sheet. In fact the lighting of a scene against back (or front) projection is very much restricted by the intensity of the light on the screen.

It is also unwise to track a camera any distance in or out in front of a back (or front) projection screen, or for that matter of a perspective photograph or drawn backing. As the camera moves the size of the foreground subjects will change out of all credible relation to the static perspective of the background picture.

This illustrates a typical back-projection set-up, with good and bad camera positions:

Fig. 17

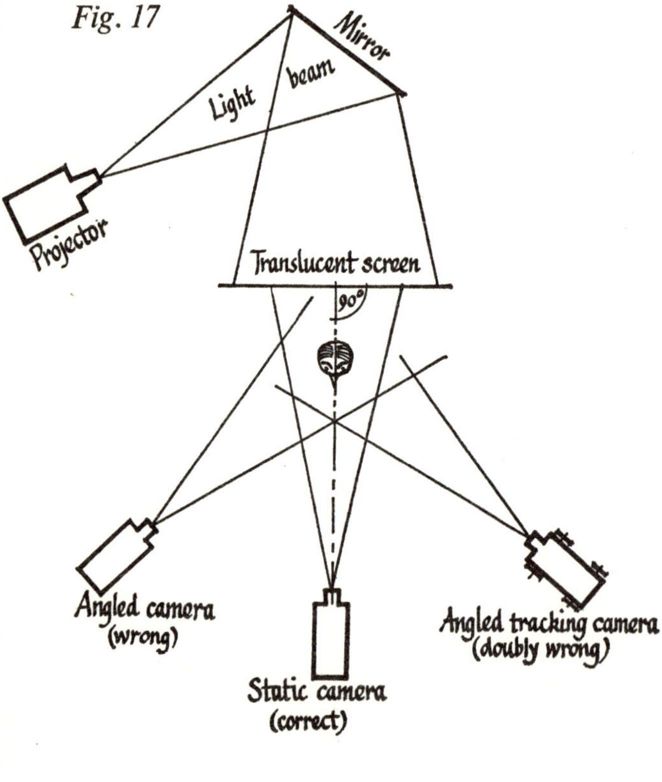

30e. Periscopes and mirrors.
Very low or very high angle shots can be obtained even from pedestal-mounted cameras if a periscope is fixed in front of the lens or if a mirror is positioned so that a scene can be shot through it. The limitation placed by these devices on the movement of the camera and the flexibility of the scene is obvious and they should only be used where unavoidable, particularly as they take a considerable time to line up.

30f. Prompting devices.
Devices attached to the front of cameras so that a speaker can read his lines from a reflection of them in front of the lens can be hired, complete with operator, from certain firms and are much used in factual programmes. They are a very real help in the televising of prepared statements, or in providing impromptu speakers with subject headings, but should not be used when an effect of true spontaneity is required except with highly experienced professional commentators, and never—except in the direst emergency—by an actor in vision in a dramatized programme. They cannot always be fitted in front of a zoom lens.

30g. Inlay and overlay.
Some studios are equipped with apparatus which enables part of a picture from one source (e.g. Camera 1) to be superimposed upon another (e.g. from Camera 2). Unlike the 'mixing' effect which appears in an electronic or optical dissolve between two pictures, and in which one image is 'ghosted' over the other, inlay and overlay give complete separation between two solid images so that, for instance, a background from one camera does not shine through a foreground figure from the other. In overlay this effect can be extended to moving objects.

Both inlay and overlay take some time to line up and their effect is not entirely natural. They are seldom used in drama production but can be useful in variety and magazine programmes, particularly in providing a wide selection of ways of 'wiping' from one picture to another, and in superimposing lettering. (In the last case they may be used in conjunction with slide-projectors—devices which make possible the high-speed cutting of captions in pre-arranged order.)

Colour Separation Overlay (C.S.O.), also known as Chromakey, is now common and operates on a specific triggering colour—often blue, since this colour is not found much in flesh tones. It is then necessary to avoid blue dresses, scenery, etc. It should be stressed that both foreground and background cameras should be strongly locked off for most C.S.O. effects.

30h. Rostra.

Platforms (or 'rostra') can be built in studios or on outside locations and either the performers or the camera or both placed on them. Naturally they place severe restrictions on movement and a good deal of time has to be allowed for their erection. When using them you will need to estimate very carefully the top limits of your sets: a pedestal-mounted camera on the studio floor can only look up at people on platforms and as a tall man on a four-foot rostrum is ten feet high you are likely to shoot off the top of the scenery and into the lights in the attempt to show his face.

For all that, a variety of floor levels and the use of higher and lower angle shots will often add interest to the camera work, provided always that the effect is not obtrusive or distracting.

In fact, you should always ask yourself when choosing or using any kind of equipment: Do I really need it in order to reveal the subject and purpose of the programme? For the success of your programme does not depend upon the sophistication of the equipment, but on how well you convey its meaning.

RECORDING AND FILMING

31. A television programme may be transmitted 'live'—i.e. seen by viewers as it actually happens—or it may be recorded in advance, either wholly or in part. It may also be transmitted live and recorded at the same time for future repeats. It may be recorded on film or on magnetic tape or both, or some parts of it may be pre-recorded on one and some on the other. Parts of it may also be pre-shot with ordinary film cameras and the film either played into the programme or cut into a filmed recording.

In the early days of television most programmes were live, but recording standards and facilities have improved so much that at present some eighty per cent of television programmes made in the United Kingdom are recorded in one way or another. A complete introduction to recording and filming techniques would require another book; all that can be attempted here is to give the beginner a general idea of current practice and of some of the possibilities and pitfalls facing him.

31a. Video-tape recording.

A complete picture and sound record of a television programme can be registered on a single roll of special wide magnetic tape and the basic principles of recording, wiping and cutting are the same as for ordinary sound recording tape. The reproduction is very good and a producer can accept a video-tape recording as 'as good as' a live broadcast, provided it is going to be transmitted on the same electronic network.

However, it cannot be shown in this form on any other system. If recorded on a 405 line system it cannot be broadcast on 625 lines without either going through a piece of equipment known as a standards converter or being rerecorded on to film.

In either case there will be a loss of definition. When it is known beforehand that a programme is required for another network (e.g. for overseas sale) it may be recorded directly on to film, and the producer or director will have to make allowances for this. (See 31c, below.)

If you pre-record a section of a programme (a scene in a play, an item for a magazine, a turn for a variety show) on video-tape, you can arrange to have it played back into the studio so that it takes its place in the rest of the programme. Or, if the rest of the production is to be recorded on the same video-tape system, you have the alternative of editing the pre-recorded section into the final recording afterwards; but remember that this cannot be done if the two recordings are on different systems—e.g. one on video-tape and one on film.

31b. Video-tape editing.

Video-tape can be cut and joined together again, so that sections of a programme can be deleted or rearranged after recording, but as the sound is recorded a fraction of a second ahead of the picture (nine and a half inches earlier on the tape) allowance has to be made for this, otherwise there will be gaps or jumps in the sound on the edited version. Your technical supervisor will be able to advise you how to do it.

Video tape can also be edited electronically with a high degree of precision. This process is becoming more and more common.

The advantages of pre-recording and editing programmes are obvious: you can pause between scenes to rearrange the studio or allow performers to change costumes; you can do a section of a programme again if things go severely wrong; or you can compile a programme over a period of days or more (such as for the actuality coverage of a protracted public event).

Some directors are tempted to use it as a form of film-making, stopping the recording between almost every shot and lining up each scene as in a film studio, but this is generally frowned on as being uneconomic. Your programmes will also lose spontaneity and flow if you stop and start them too often.

31c. Film recording and editing.

Programmes can be recorded in similar ways on 35mm. or 16mm. film. Of these 35mm. gives the better quality (though not always up to the standard of video-tape recordings), while 16mm. is more generally marketable internationally. In both cases the sound is usually recorded simultaneously on separate magnetic tape.

Film recording with a separate magnetic sound track has two advantages over video-tape. It can be transmitted on electronic systems other than that by which it was shot, and it can be edited 'in parallel sync': i.e. the picture and sound can be cut at the same point without pauses or jerks in the edited track.

If your programme is being recorded on film you should watch for two things: first, you should avoid fast pans or movements across screen which may blur in reproduction; secondly, remember that film recording has a 'cropping' effect—that is, about three per cent of the height or width of the picture will be cut off from each edge if it is recorded and transmitted on film. Thus any action or detail at the edge of the screen will be lost to the viewer. The best way to deal with this is to keep your important action in the middle of the screen, where it should be anyway.

If many cuts are made in the print of a film recording, it will be necessary to cut the negative to match and have a new print made from that. As this not only costs money but takes a day or two to do, you must make sure that there is sufficient time for it between the recording and transmission of your programme.

31d. How to use recordings in the studio.

If you want to include pre-recorded inserts in your programme you will naturally have to make sure that machines are available to run tape or film at the time you need them, and also that the quality and style of the inserts match the rest of the production. Remember, too, to watch continuity of action and costumes: to take an extreme example, imagine the effect of an actor or speaker who was wearing a moustache when the insert was shot but has shaved it off by the time you come to do the rest of the programme.

When playing film or video-tape inserts into a programme, bear in mind that it takes about ten seconds for the machines which are feeding them to you to run up to their proper speed: you must therefore cue them ten seconds before you need them. This will require practice, for which you should allow time in rehearsal.

You can then cut between your cameras and the pre-recorded visuals, or fade out one and fade in the other. You cannot dissolve between one and the other unless the equipment from which the recording comes is electrically 'locked' with the electronic equipment of the studio (or outside broadcast unit) by one of two processes known as 'gen.-lock' or 'slave-lock'. As this is a complex arrangement you must check with your technical supervisor in advance if you want to 'mix' between your cameras and the pre-recordings.

31e. Film inserts in programmes.

Sections of film already shot with ordinary film cameras may be included in a television programme in the same way as film recordings. The same problems of matching and continuity apply—in fact you need to be even more careful to see that such things as the lighting and the action of the film are suitable for the programme, as the filming may have been done in a number of different locations under widely varied conditions.

The basic grammar of camera direction and picture composition holds good for films as well as for television, but the means of achieving the same ends are in many ways different. Filming is a complicated and separate craft and there is no room in this book to deal with it adequately. Even when you have mastered the fundamentals of directing television cameras you will still not be a qualified film-maker. The best advice to the beginner in this field is: tell your film technicians what results you want and follow their advice in achieving them. You will learn from their experience.

Above all, make sure that the tempo and style of the film inserts do not fight against the general character of your programme. Film editing for the cinema tends to have a different rhythm from television cutting; make sure it does not conflict with the way you are directing in the studio.

REMEMBER—

Whether your programme is to be 'live' or recorded, the sole purpose of your technique, and of the equipment you use, is to produce a series of images which, with their accompanying sound, will please, satisfy and enlighten the viewer. The basic grammar of television production exists to help you to do this, and should not lightly be disregarded for the sake of an effect. Never forget that the simplest ways are often the best, and that the subject of your programme should be the master and your technique its servant.

GLOSSARY OF TERMS
(For the completely uninitiated)

Angle. (Camera angle.) The line along which the camera looks at the subject. The angle of the camera to the subject.
Angle of acceptance. The angle governing the breadth of vision of the lens. Some lenses have a narrow angle of acceptance which enables them to take a close shot from a great distance. Some have a wide angle of acceptance which enables them to take a long-shot from a small distance. Because the picture has an aspect ratio of 4 × 3, is wider than it is high, the horizontal angle of acceptance of a lens is greater than its vertical angle of acceptance. However, a lens is commonly known by its horizontal angle of acceptance.

Fig. 18

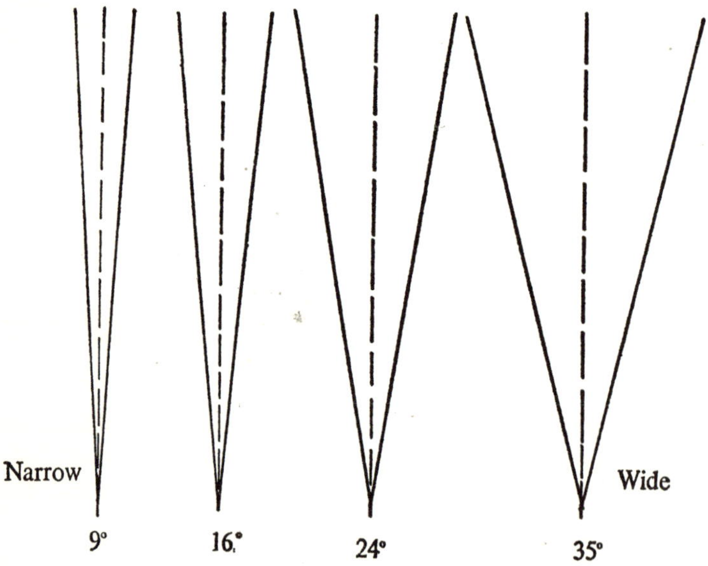

Fig. 18 is a diagram showing the horizontal angles of acceptance of the four lenses most commonly used on television for studio work. Both narrower and wider angle lenses exist and can be used.

Aspect ratio. The ratio of the horizontal edge of the frame to the vertical edge of the frame. The frame of the television screen is 4 × 3. If it is sixteen inches wide, it is twelve inches high. If it is twenty inches wide, it is fifteen inches high.

Back projection. A method of providing scenic settings by projecting them, either from a slide or from film, on to the back of a translucent screen.

Background. That part of the picture that is farthest from the camera.

Boom. The sound boom. It is a device which holds the microphone close to the artists but always out of shot. It has a retractable and extensible, telescopic arm which is pivoted at one end. By this means the microphone can be made to follow the artist wherever he may move. It also has a device for turning the microphone so that one artist may be 'favoured' more than another.

Boom Operator. The person who operates the sound boom which carries the microphone.

Call-boy. The person who fetches the artists or performers when they are required.

Cameraman. The person who operates the camera, keeps it in focus and frames the subject.

Camera script. A completed, final script of a television performance giving the dialogue on the right-hand side of the page and the shots and camera movements and other technical details on the left.

Caption. Written or printed matter that is put in front of a camera.

Caption Scanner. A simple, fixed-focus camera for transmitting written or printed captions can be transmitted without using a complete camera.

Close shot. A picture in which the subject appears large in the frame. It probably cuts his chest.

Close-up. A close shot.

Colour Separation Overlay. See page 76.

Commercial. Advertising matter transmitted between programmes or in natural breaks in programmes.

Commercial Break. The natural break in a programme, such as between the acts of a play, during which advertising matter is transmitted.

Complementary angle. See rule 21.

Control room. (Gallery.) The room from which the director governs and controls the rehearsals and performance. Most of the personnel working in the studio wear head-phones and he can speak to them and give them instructions.

Crab. To move the camera sideways.

Crane. Camera crane. A very large type of 'dolly'. Both the camera and the cameraman are mounted on the end of a very long, counter-weighted crane-arm which enables them to be swung to the left or the right, up or down. There are several sizes of crane. They are tracked in and out, electrically. They are large, cumbersome and difficult to get quickly from set to set but, once in position, they make for great fluidity and mobility of camera movement and will give greater height than any other mounting.

Credit. An acknowledgement by name, either in sound or vision, of an artist or performer or other person who has contributed to a programme.

Cue. A signal to start to do something. This may be a prearranged word or a sound or a light or a hand signal. All cameras have cue-lights on them to show when they are operating 'On the air'.

Cut. An instantaneous switch from one camera to another. It should always be used when the action is continuous in time.

Cut-key. A button which can be pressed by the prompter, momentarily to cut all studio sound while he prompts an artist.

Depth of focus. See Hints and Tips, 16e.

Director. The person responsible for the organization, artistic interpretation and the presentation of a programme.

Dissolve. Also called a mix. A gradual merging of one picture into another till the first picture has disappeared and only the change from one lens to another. The normal lens turret carries four different lenses. (See Angle of acceptance.)

Dolly. A wheeled truck on which a camera may be mounted. By means of the dolly the camera is enabled to 'track', be pushed towards or pulled away from the subject.

Eye-line. The line along which a person is looking. The direction of his look.

Favour. To see more of one person's face than that of another person in the same shot.

Fifty-fifty two-shot. A picture containing two persons only, in which the camera favours both equally. That is to say, that it does not see more of the face of one than of the other.

Focal length. The distance between the optical centre of a lens and the plane at which an image of an object at infinity is formed.

Fold Back. To play music or sound effects from tape, disc, etc. and feed them back into the studio so that they can be heard by the artists and performers.

Foreground. That part of the picture that is nearest to the camera.

Gram Operator. The person who plays music or sound effects from tape or disc in the sound control room.

Head-room. The distance between the top of the head of a subject and the top edge of the frame.

Individual shot. A picture of one person only.

Inlay. A method by which the picture from a television camera is inserted into an area of a scene portrayed by a second camera. The outline and magnitude of this area being determined by a third device known as a silhouette generator. (See Overlay.)

Jib. The crane-like arm of a dolly on which the camera is mounted. This arm is fixed to a turntable which can be revolved so that the camera can be moved to the left or the right. This movement is known as 'jibbing'. The jib can also be raised and lowered.

Left, right of frame. Camera and audience's left and right. Not subject's left and right.

Lens turret. A device on the front of the camera operated from the back of the camera by which the cameraman can quickly change from one lens to another. The normal lens turret carries four different lenses. (See Angle of acceptance.)

Lighting Supervisor. The technician who is in charge of the lighting of a scene or a production in both the technical and artistic aspects.

'Live' television. Television, the action of which is taking place at the moment at which the viewer is watching it. Television that is not recorded either on film or magnetic tape (Video-Tape.)

Long shot. A picture in which the subject is relatively small in the frame, i.e. relatively far from the camera.

Matched shots. See rule 21.

Mid-shot. Something between a long-shot and a close-up. It probably cuts the subject at approximately the knees.

Mix. Also known as the dissolve. See dissolve.

Monitor. A television set in the studio or its control gallery, giving the picture emanating from that studio, or other chosen source.

Mute. Silent. Picture on film that has no sound track.

Narrow angle lens. A lens with a small angle of acceptance.

Natural perspective. A perspective that is similar to that of the human eye.

On the Air. Being actually transmitted. Normally, except in the case of a superimposition, out of three or four cameras only one at a time is being transmitted, is 'On the air'. The others are said to be 'Off the air'. It is while cameras are 'Off the air' that they are moved to new positions and changes of lens take place.

Overlay. Colour Separation Overlay (CSO), also known as Chromakey. A method of separating the foreground from the background so that the background picture can be selected at will. Blue is normally the colour used to 'key' the background. There must be no blue in the foreground picture if the 'keying' is to work correctly.

Over-shoulder shot. A picture of someone with the back of another person's head and shoulder in the foreground.

Pan. To turn the camera to left or right about its vertical axis. It is the verb form and abbreviation of 'panoramic shot'.

Pedestal. Camera pedestal. A camera-mounting which is vertical and smaller than a dolly. In the past these were less mobile and less adapted to tracking. The modern ones, however, are excellent in this respect and are superseding the dolly. On the modern pedestal it is also possible to raise and lower the camera, in vision, at the touch of a finger.

Preview. (*verb*) To examine a picture before it is switched over to transmission. (*noun*) A monitor showing such a picture.

Producer. This is sometimes used as a synonym for 'director'. Usually the producer is responsible for the overall planning in a broad sense for a series or group of programmes while the director is responsible for the detailed control of one particular programme.

Production Assistant (P.A.). The director's personal assistant on a programme.

Props. Theatrical properties. Objects used, and seen by the audience, in the course of a production. 'Personal Props' are those actually carried by an actor or performer.

Roller caption. A strip of written or printed matter which is passed in front of the camera by means of rollers.

Rostrum. A platform built in the studio as part of the set.

Set. (Verb.) To place in position.

Set. (Noun.) A complete arrangement of scenery representing one place, erected in the studio. A production may have several sets representing different places.

Shot. A picture from a camera.

Sound mixer. The technician who balances the output of sound from the various microphones, the record players and the tape machines and mixes them in their right proportion and fades them in and out when required. He is responsible for the entire sound output of a given studio.

Strike. To remove (an object) from a set.

Strobing. A streaking, juddering effect caused by the camera and the subject moving too fast in relation to one another.

Studio floor manager. The person who is responsible for the organization and discipline of the studio for a particular production. By means of head-phones, he listens to the director in the control gallery and relays his instructions to those who are not wearing head-phones. He is the director's representative in the studio while the director, himself, is in the control gallery.

Subject. The person(s) or thing(s) at which the camera is looking.
Superimpose. To place the picture from one camera on top of the picture from another and to transmit them both at once. This gives a 'ghost' effect with one picture showing through the other. It is most generally used for captions when the lettering is superimposed on a background picture.
Tape. Magnetic tape upon which sound is recorded and from which it can be replayed.
Telecinė. Film projection used in television. This may constitute an entire programme or part of it.
Three-shot. A picture in which three persons only are just contained in the frame.
Tilt. To tip the camera upwards or downwards.
Track. To wheel the camera forwards or backwards, towards or away from the subject. The track in gives a closer, tighter shot. The track out (i.e. back) gives a longer, wider shot.
Two-shot. A picture in which only two persons are just contained in the frame.
Video Tape. Magnetic tape upon which both vision and sound are recorded and from which they can both be transmitted.
Viewfinder. On a television camera, this takes the form of a small screen, shaded by a hood, on the back of the camera. The cameraman looks into this in order to compose and frame his picture and to keep it in focus. There is, also, the director's pocket viewfinder. This is a small optical device which enables him, at rehearsals, to see exactly what the camera will see on a given lens at a given distance.
Vision mixer. The operator who, according to the script and the instructions of the producer or director, cuts or dissolves (mixes) between one camera and another and fades cameras in and out.
Whip pan. See rule 17.
Wide angle lens. A lens with a large angle of acceptance.
Wipe: To change shots by using an inlay device.
Zoom lens. A lens device which gives an infinitely variable angle of acceptance.

ACCEPTED ABBREVIATIONS (on scripts)

A.B. = As before. It is used in connexion with a shot from a camera when that camera has exactly the same shot as it had immediately before.

BCU. = Big close-up. Head or part of the head. Sometimes written VCS. (very close shot).

B.P. = Still or moving back-projection.

Cam. = Camera.

CMS. = Close medium shot. Cutting the subject below the waist. This can be a tight three-shot.

CU. = Close-up. Head and shoulders. Sometimes written CS., i.e. close shot.

F.R. = Film recording.

F/X = Effects (sound) recorded on tape or disc.

Grams. = Music or sound effects from gramophone records.

LS. = Long shot. The figure occupying approximately half the height of the screen.

MCU. = Medium close-up. Cutting the subject just above the waist. This can be a two-shot. Sometimes written MCS., i.e. medium close shot.

Mic. = Microphone.

MLS. = Medium long shot. A tight, full-length figure.

MS. = Medium shot. Cutting the subject at the knees.

O.S. = Overscene. Applied to speech over a picture in which the speaker does not appear. Also used in connexion with music or sound effects when their source is not visible.

O.O.V. = Out of vision. Speech, music or effects which are heard when their source is unseen.

O.S.V. = Off-screen voice. The voice of a person who is heard but not seen.

S.O.F. = Sound on film (effects, music or speech recorded on film track).

Spot F/X = Sound effects made physically in the studio and not recorded on tape or disc.

TK. = Teleciné, i.e. film or film recording played on a teleciné machine.

TX. = Transmission.

VLS. = Very long shot. The full set.

VTR. = Video-tape recording.